IMAGES
of America

BURLINGTON FIREFIGHTING

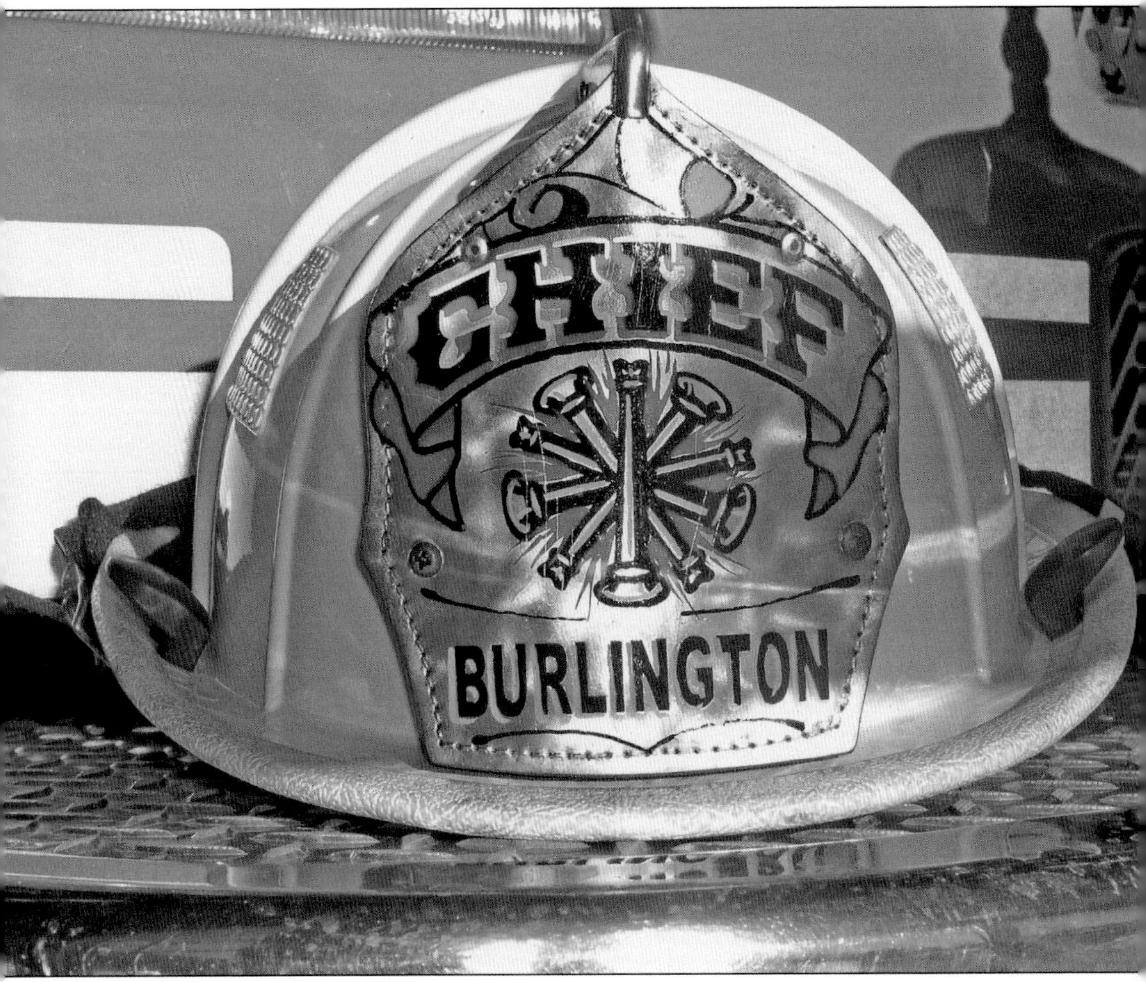

The fire chief's hat is a symbol of the fire department, well recognized worldwide. Today's helmets appear unchanged, but in essence they have improved in design. The improved linings provide comfort as well as safety. Traditionally the outside of the helmet has been made of leather. However, man-made materials are now being used as well, keeping comfort and safety in mind. (Courtesy of Burlington Fire Department.)

On the cover: Please see page 93. (Courtesy of Burlington archives.)

IMAGES
of America

BURLINGTON FIREFIGHTING

Toni Faria and
the Burlington Historical Commission,
Foreword by Chief Lee Callahan

ARCADIA
PUBLISHING

ISBN 978-0-7385-5748-9

Published by Arcadia Publishing
Charleston SC, Chicago IL, Portsmouth NH, San Francisco CA

Printed in the United States of America

Library of Congress Catalog Card Number: 2007938713

For all general information contact Arcadia Publishing at:
Telephone 843-853-2070
Fax 843-853-0044
E-mail sales@arcadiapublishing.com
For customer service and orders:
Toll-Free 1-888-313-2665

Visit us on the Internet at www.arcadiapublishing.com

For all Burlington firefighters past and present

Contents

ACKNOWLEDGMENTS

This book was a cooperative effort. As a novice, I needed help gathering information and pictures. Thankfully people came forward with information, stories, and pictures to make the book possible.

I would like to thank the following people who made the book possible. Retired fire chief Herbert Crawford was the first to come forward with information and photographs. He shared his knowledge, memory, and love of the department, as well as photographs. Robert Washburn, a retired firefighter from Lexington, shared his research and photographs. Retired firefighter Albert Fay shared his collection of photographs, which made it possible to complete the book.

Burlington fire chief Lee Callahan and assistant chief Steven Yetman were exceptionally cooperative, as were Capt. Michael Patterson, dispatcher Elaine Carpenter, Capt. Scott Carpenter, and the office staff, JoAnne Arbing and Mary Fay.

Lt. Robert Paul and Lt. Peter McAnespie and firefighters Brendan Miccichi, Kevin Pollicelli, Ernest Covino, and Gerald Hanafin helped with information. The Burlington Firefighters Local 2313 and Pres. Edgar McLean supported the efforts to have the book printed.

Retired firefighters William Callahan, George Clark, George Creamer, Thomas Corbett, James Dannaher, William Devereaux, Joseph Doran, Albert Fay, James Fay, Robert Ferro, James Hapenney, Paul McLaughlin, Lawrence Rice, Donald Robinson, Robert Sullivan, Paul Thibault, and Major Travers provided photographs and information. Steve Duke of the auxiliary fire department and water treatment plant manager William Keene provided information, as did Robert Clougherty and Wayne Higden.

The Burlington Cultural Council provided a grant, and the historical commission supported the creation of the book. Commission members Eldrine Emerson and Joyce Fay helped in obtaining photographs.

I thank Michael Tredeau of the commission for the many hours of scanning and printing of images. I am grateful for the time and expertise he contributed to help complete the book.

Burlington residents Margaret Butler, Joyce and Robert Carpenter, Shirley Mason Chepulis, Geraldine Gaffey, and Nellie Russo provided details and photographs, as did Judith Jango Cohen, John White, and the Woburn Daily Times. Town of Burlington archivist Dan McCormack provided the early photographs of people and sites. The staff of the selectman's office also was supportive.

Erin Rocha of Arcadia Publishing was helpful and patient throughout the process.

Lastly, I thank my children for their encouragement and my late husband, Lt. Lawrence Faria, who shared with me his love of the fire service and inspired the creation of the book.

FOREWORD

For many people today, their firefighters have always been there to rescue them, their loved ones, and their property. In reality, that is not so in Burlington. This book starts at the beginning of the Burlington Fire Department and brings the reader up to the present.

More than 100 years ago, the farmers, tradesmen, and merchants, as well as residents, of Burlington responded to another's call of "Fire!" The start of the fire service in Burlington was disorganized but intent on saving lives and property, just as the organized fire service of today. From the completely volunteer efforts of the Skeltons and others to the call firefighters of the middle of the 20th century to the professional firefighters of today, the commitment of providing life safety has been paramount to each of those involved. The training and technology has changed, but the core mission has remained the same: lives and property.

The following pages will give a glimpse into the history of the Burlington Fire Department: where we started, how the journey paths changed, and where we are today as an emergency service. The journey is not over. The Burlington Fire Department now provides public fire safety education, code and regulation enforcement, hazmat response, technical rescue services, home and commercial building inspections, motor vehicle crash response, emergency medical aid treatment and transport, emergency communication services, and a host of other public safety–related services, as well as fire suppression, from whence it all started. From the initial \$185.35 spent on municipal fire costs in 1915 to an annual budget of more than \$5.3 million today, the fire department has grown along with the town. The history of the Burlington Fire Department should encourage the reader to appreciate even more the emergency services they enjoy today. I hope you enjoy reading of our history and looking at the photographs captured by various photographers over these many years. They define the tradition of the fire service in Burlington.

—Fire Chief Lee Callahan

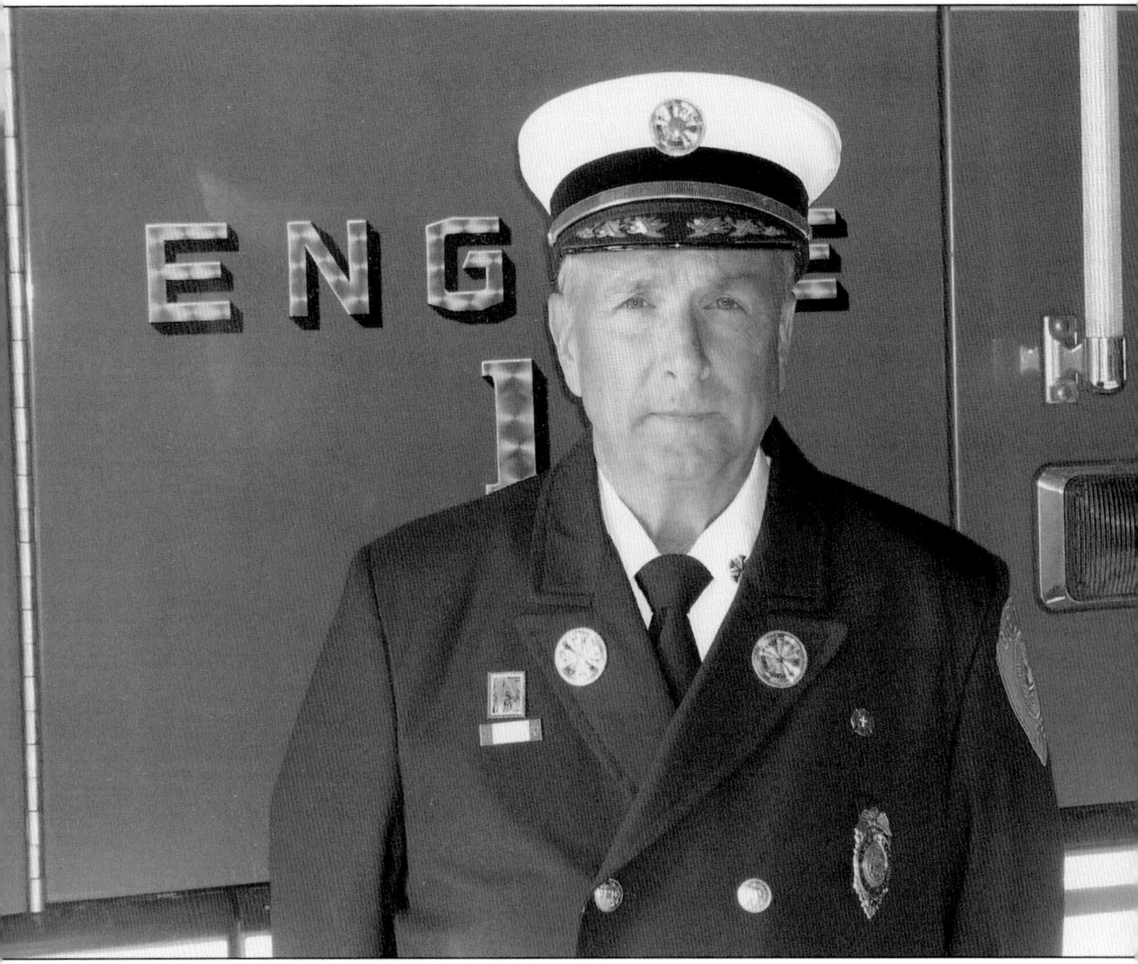

Burlington fire chief Lee Callahan was appointed in 2004 and continues to lead his trained and knowledgeable firefighters to meet the demands and challenges of the present-day fires, rescues, and emergencies. (Courtesy of Lee Callahan.)

INTRODUCTION

Burlington was settled about 1642 and at that time was a part of Woburn. In 1799, the town was incorporated, and by 1800, the census counted 534 inhabitants. Those 534 people were responsible for protecting their own property from fires of any sort. As the town grew, farms and homes became bigger and fires became a serious problem.

In the past, historically important homes were lost to fires. The 1799 Wood Tavern in Burlington Center, where the town was incorporated in 1799, was razed in 1960 after a fire. The James Reed house where British prisoners were held in 1799 was lost to a fire in 1951. The 1700s Sewall House, featured on the town seal, was destroyed in 1897 by a fire. John Hancock and Samuel Adams sought refuge there in 1775. In 1937, the Cummings three-story mansion burned to the ground. It was the largest farm in Burlington with 294 acres of land. More recently, the Francis Wyman house built in 1730 on the original 1666 site burned in November 1996.

Farmers were using simple firefighting methods. It was not until 1900 that a volunteer group of men began to organize. By 1907, the new forest/tree wardens, Walter Skelton and Edward Bennett, who was the first forest/tree warden in 1903, were paid for "watching fires." That year in the annual report, it was noted that five fires burned 115 acres of wood with an estimated damage of $725. The fire alarm at that time consisted of Skelton ordering that the church bells be rung twice, which alerted the volunteers to assist in putting out the fire. In 1912, fire alarm gongs were installed at the Union School. Not until the 1920 annual report were fire trucks mentioned for the first time, and then, simply for the transportation of men and equipment. In spite of firefighting efforts, by 1926, 10 of 16 burning buildings burned to the ground. Three years later, the town voted $1,000 for fire protection but voted down a study on water supply and a new fire station. A fire that threatened the whole southern end of town in 1932 required help from Arlington, Lexington, and Waltham fire departments and a state pump from Danvers. The seriousness of the situation prompted the town to vote on the purchase of a pumper and new hose. In the next 15 years, more equipment was acquired, and in 1950, the first ambulance was purchased for $600.

As the town began its rapid growth in 1949, the fire department kept up with the progress by replacing old equipment with modern equipment. It became a full-time department in 1955, with Herbert Crawford chosen as the fire chief. Firefighting entered a new phase of education and training to keep up with the economic and social changes in the town.

The Burlington Mall was built in 1967, followed by the Lahey Clinic, built in 1970. Each facility brought new challenges and training to the department. The daytime population increased and resulted in an increase in motor vehicle accidents and emergencies.

Under the leadership of Chief Herbert Crawford, Burlington firefighters once again persevered and met the new demands. The continuing changes in Burlington demanded a more trained and educated department. The fire chiefs that followed Chief Crawford, after his retirement in 1984, each brought their particular expertise to the position. Acting chief Lawrence Rice was chief from 1984 to 1986. In 1987, Fire Chief Paul Thibault was appointed and served until his retirement in 2004. The present fire chief, Lee Callahan, was appointed in 2004. Thanks to the leadership provided by the officers and the caliber of the firefighters who are drawn to the profession, we have a professional department that has created a better quality of life in Burlington.

One

The Early Years

Burlington was a farming community from the settlement years of 1642 to 1950. The farmers and residents became the first firefighters to protect their farms, homes, barns, and livestock. The ability to extinguish fires depended on the availability of water. Water buckets were taken to the nearest pump or stream to fill to begin the bucket brigade. This method was not always effective, and many valuable properties were lost to fires. There were no other means of fighting fires in Burlington for the first 100 years of its existence.

In 1909, the first firefighting equipment consisted of a wagon, which held 25 milk cans to carry water to the fires. The next year, residents organized a group to watch for fires. The men listed as volunteers were leaders of the community, such as John Nichols, Charles Perry, H. W. Graham, Henry Lawrence, and Horace Skelton. Dairy farming and fruit and vegetable production thrived, as well as the Reed Ham industry, Shoddy shops, and the first housing subdivision in the Winnmere section of town. Dairy and vegetable farmers such as the McIntire brothers and the Cutler, Reed, Winn, Marion, Walker, Crosby, and Bustead families joined the group of firefighters. In 1911, 75 acres of land burned.

In 1912, the governor appointed a commission to investigate the possibility of adequate water supplies for cities and towns. Not until 1931 was the town given permission to put in a water system. The work did not begin until 1949.

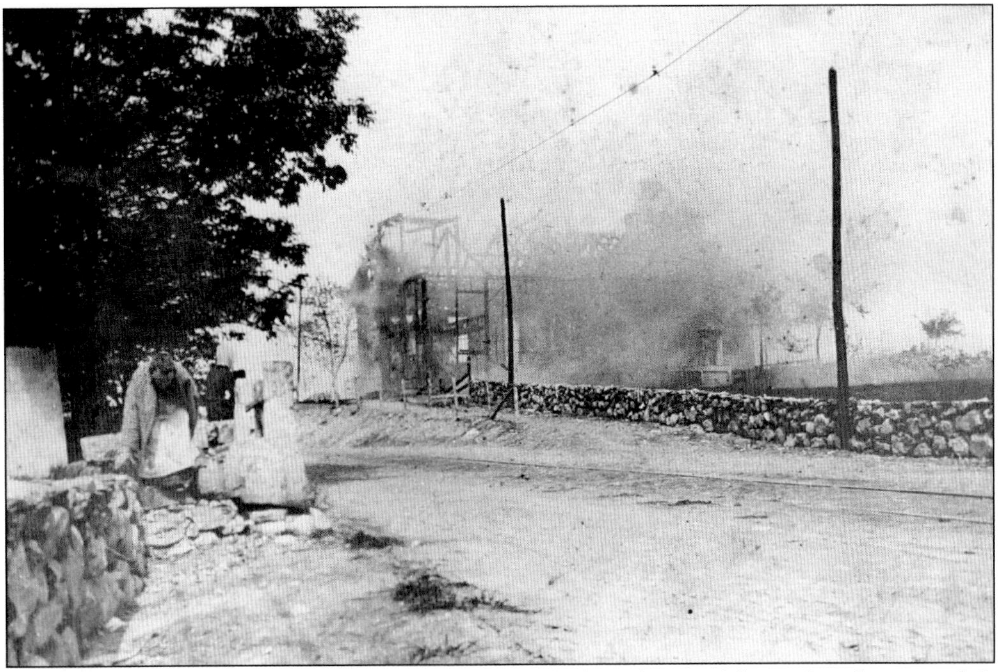

In 1902, the town house, built in 1844 across from the burial ground, burned to the ground. It resulted in the loss of all town records. Farmers were the first firefighters, making it difficult to fight and contain large fires. As shown in the photograph, there was no firefighting equipment to contain the fire or prevent it from spreading. (Courtesy of Burlington archives.)

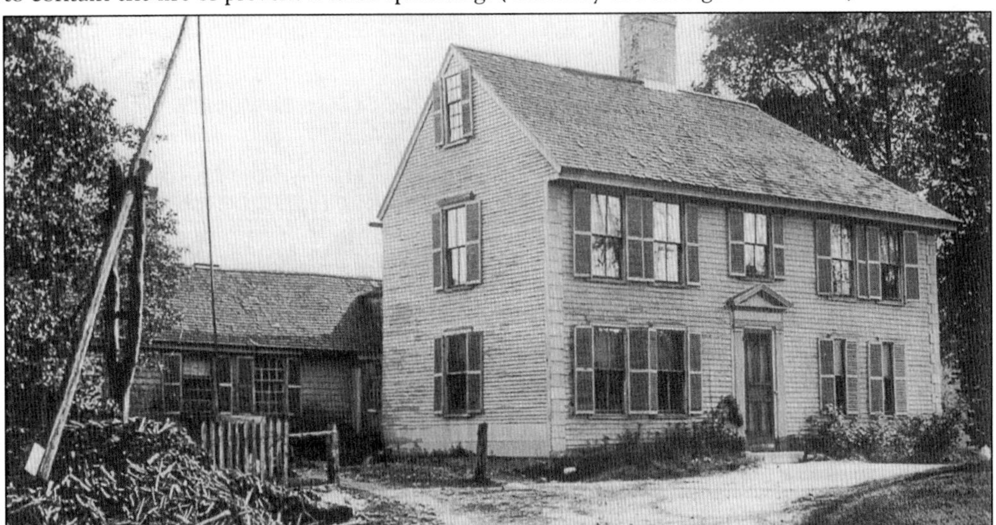

In 1897, the historic Samuel Sewall House was lost to a fire. It was the home of Burlington ministers for over 100 years. In *The History of Burlington 1640–1950*, Lotta Cavanaugh Rice Dunham gives the following information: "In 1897 when the historic Sewall mansion burned, Burlington's Fire Department consisted of everyone in town who could carry a bucket of water and fight the blaze hand to hand. There were no electric pumps to pump the water from the deep wells on the old place. Instead, windlass and crank and rope did their share." (Courtesy of Burlington archives.)

The fire bucket in this photograph was the type used by early firefighters and homeowners. When fire was sighted, everyone was summoned with their fire buckets, at least two of which hung in the hallway of every home in town. They would throw out their buckets to a neighbor or passerby, knowing they would be returned later since each leather bucket had its owner's name painted on it. The Marion Tavern had two such buckets hung just inside the front door for years. (Courtesy of Albert Fay.)

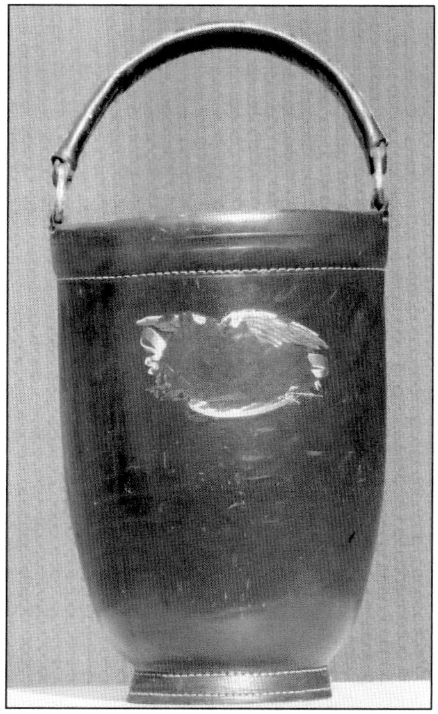

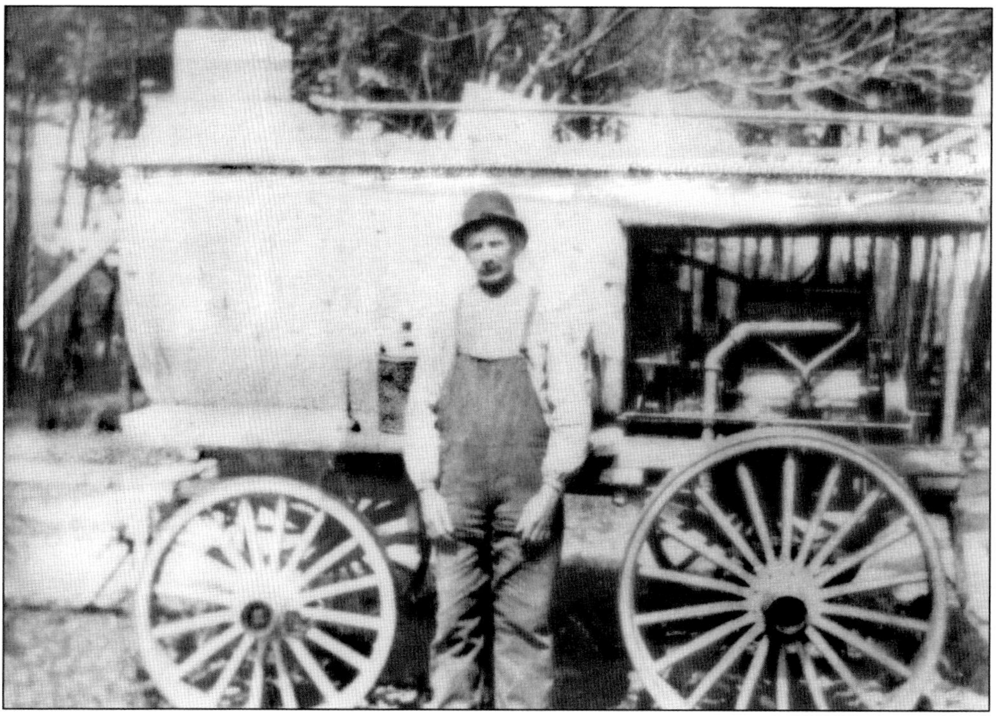

In 1907, Walter Skelton was named the first forest/tree warden and is shown in this photograph with the early spraying vehicle. His interest in firefighting was becoming evident as he began organizing men to fight fires. With the help of local farmers and residents, he soon formed a group of men watching for fires and later became the first fire chief. (Courtesy of Burlington archives.)

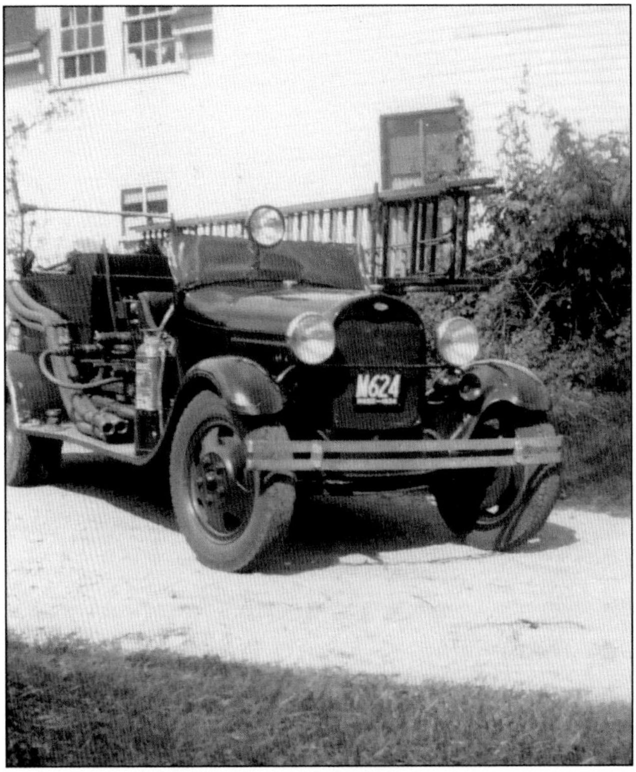

The department's first fire truck was housed in a barn across from Walter Skelton's home on Francis Wyman Road. The vehicle was a wagon, a Ford Model T with a turtleback bought at auction from H. P. Hood Company. They cut off the back, built a body with sides a foot or two high, loaded on their brooms, shovels, axes, and extinguishers, and the first fire truck was born. (Courtesy of Herbert Crawford.)

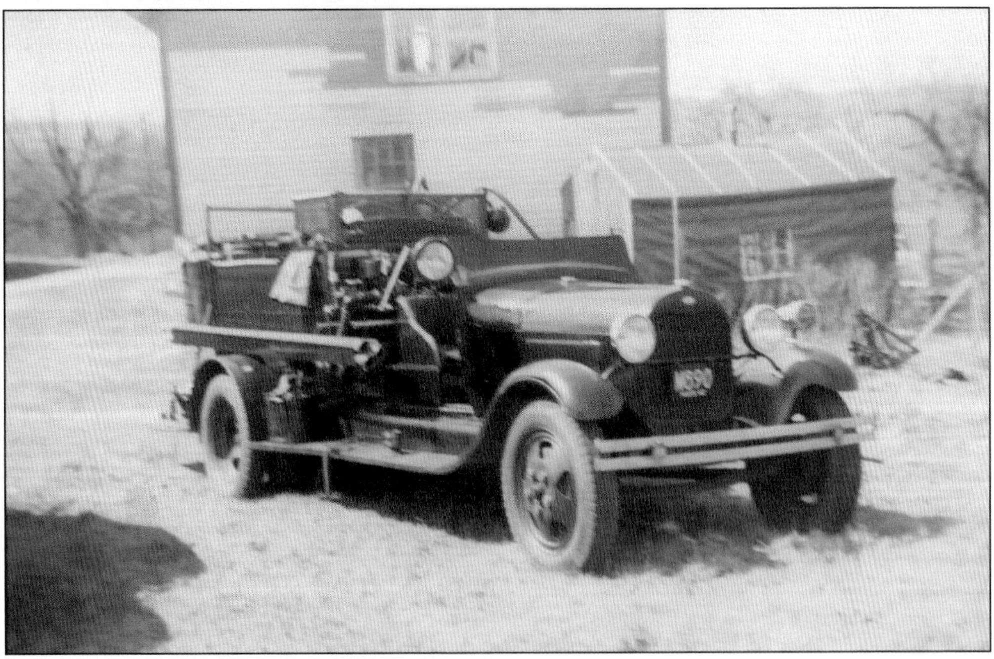

A 1927 Ford was donated to the department and turned into firefighting equipment. The men continued rebuilding vehicles as needed, gaining experience as they reconstructed the pieces. The resourceful group of volunteers became adept at reconstructing old vehicles, and soon the equipment helped contain fires. (Courtesy of Herbert Crawford.)

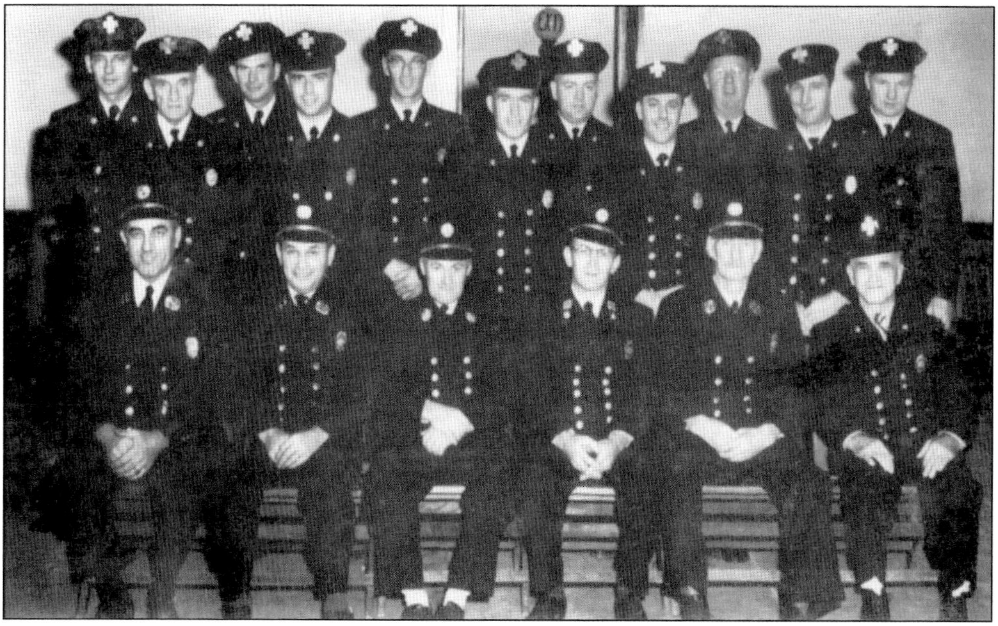

In 1938, this large group is an example of the dedicated residents who worked together to fight fires. From left to right are (first row) Capt. Sidney Brown, Capt. Robert Carpenter, Deputy Chief Charles Bunton, Chief David Skelton, Capt. Horace Skelton, and Edward Miocchi; (second row) Chester Knowles, Walter Frederick, Herbert Crawford, George Gaudet, and Louis Skelton; (third row) Donald Ellsworth, Kenneth Brown, Wilbur McIntire, Edward Howard, Allen Cronin, and Clifford Skelton. (Courtesy of Herbert Crawford.)

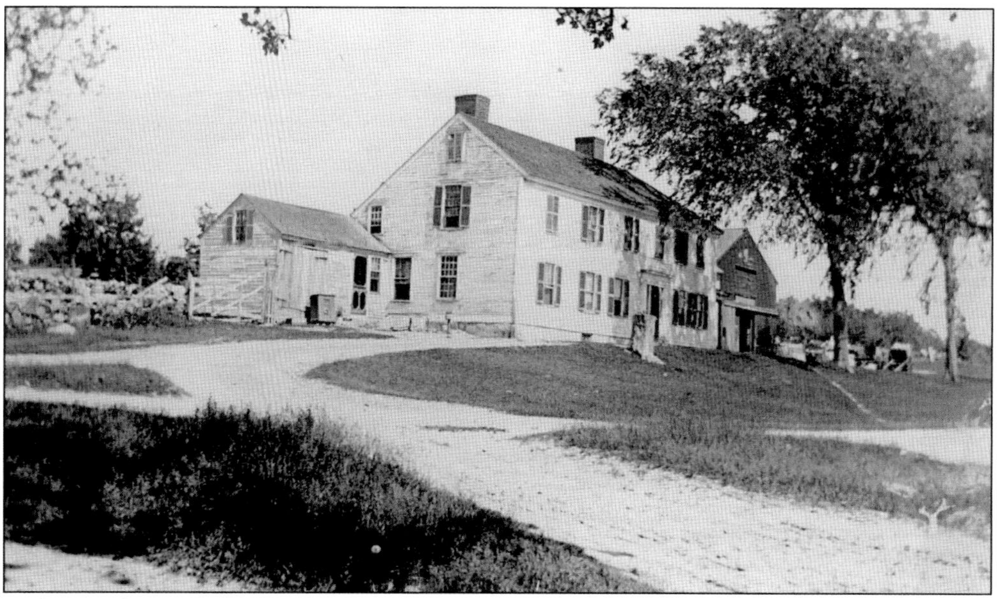

Walter and George McIntire were leading dairy farmers and owned the Marion Tavern/Grandview Farm. The property had a large rear barn and a front barn that held tools, feed, and animals. As farms became larger, the loss of a barn to fire was a constant fear, thus firefighting was important to the safety of the properties. Walter and George McIntire were part of the early volunteer group. (Courtesy of Burlington archives.)

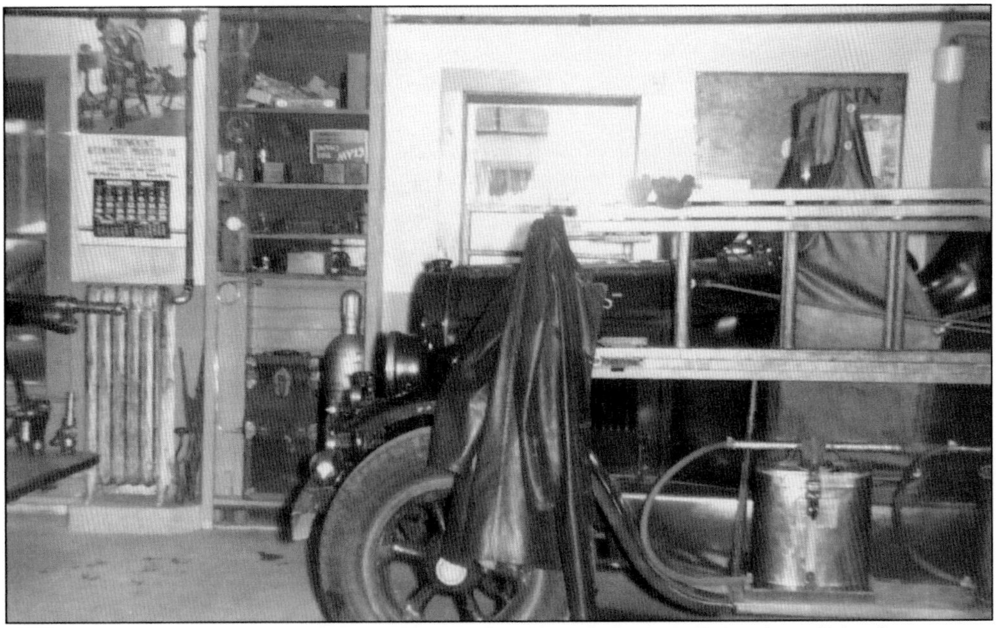

The Burlington Grange bought a new Graham Dodge chemical truck for the call fire department in 1928. The Burlington Grange, dedicated to better community life, supported firefighting by contributing $1,000 to purchase fire apparatus and fire extinguishers needed for the truck. The town provided the remainder of the money, and the new equipment was purchased. (Courtesy of Herbert Crawford.)

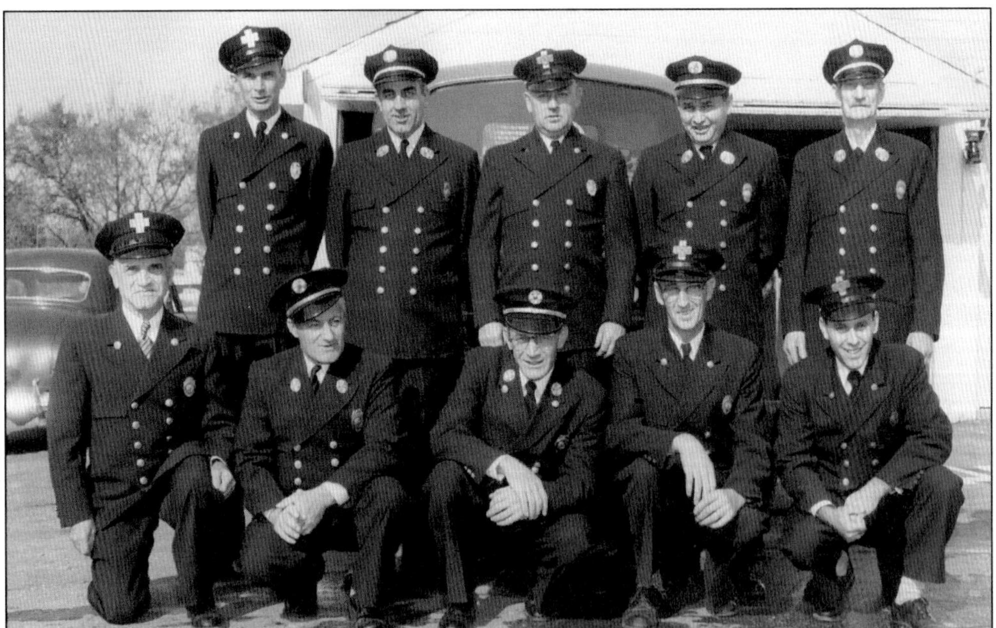

In this 1936 group photograph, the volunteers proudly pose in uniform. Pictured are, from left to right, (first row) Edward Miocchi, Charles Bunton, David Skelton, Wilbur McIntire, and Edward Gaudet; (second row) Kenneth Brown, Sidney Brown, Edward Howard, Robert Carpenter, and Horace Skelton. (Courtesy of Albert Fay.)

Completion of the annual 40- to 50-foot tower of wooden railroad ties is shown in this 1937 photograph at Simonds Park on the Fourth of July. The Burlington Men's Civic Club sponsored it. George Perkins, Maurice DeMone, Ernst Makechnie, Loren Blenkhorn, Carl Bussey, William Porter, and Chester MacDonald were members. The men proudly stand at the base of the completed tower. (Courtesy of Burlington archives.)

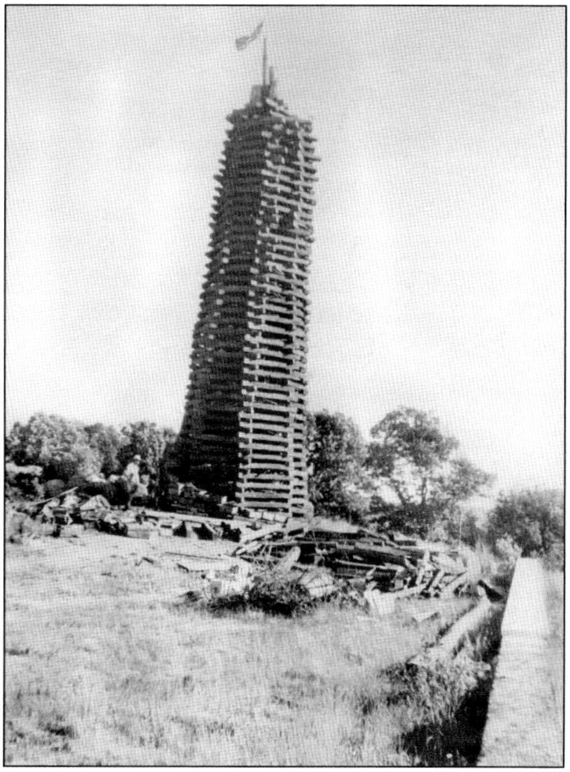

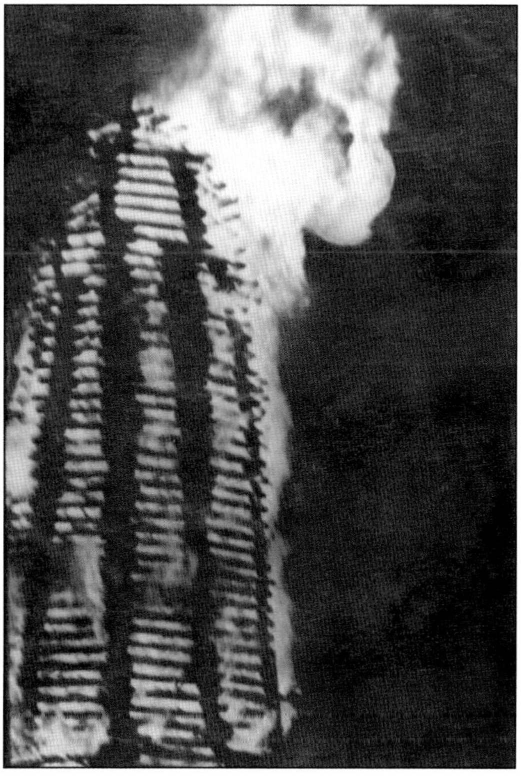

The Fourth of July was celebrated by lighting the bonfire. The last fire was held in 1940. It soon became dangerous. At one of the fires, the wooden ties collapsed, releasing the barrels that were placed on top. They tumbled off the tower and rolled down the hill, threatening to start a fire. (Courtesy of Burlington archives.)

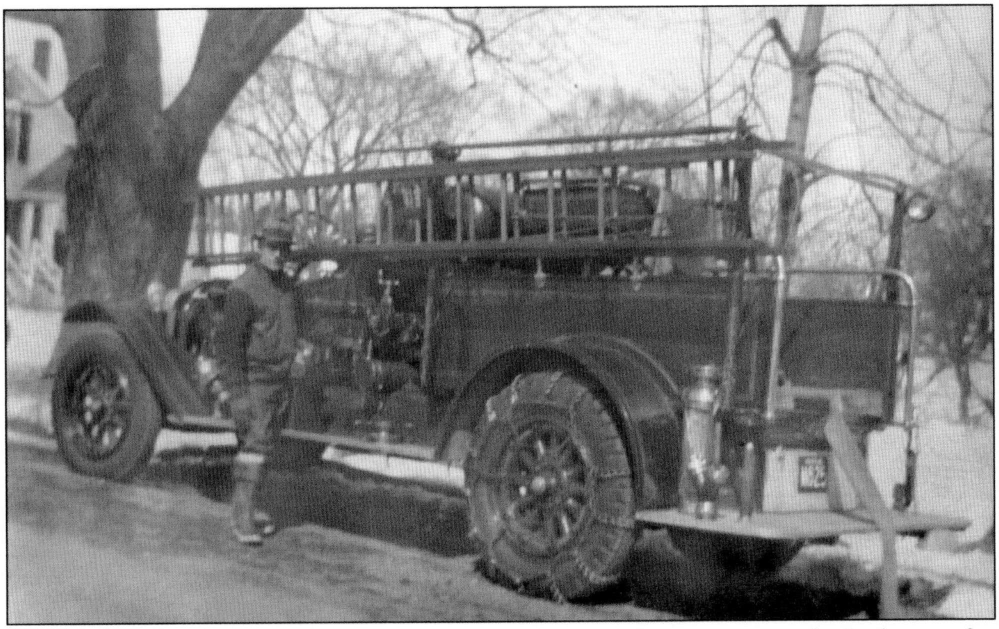

David Skelton, fire chief in 1938, is seen with an early piece of equipment. As the Burlington fire chief, Skelton followed his father's example to bring about changes to improve the department. He was untiring in his dedication to fighting fires. As a member of the Skelton family, who were early settlers of Burlington, he continued their interest in community involvement. (Courtesy of Albert Fay.)

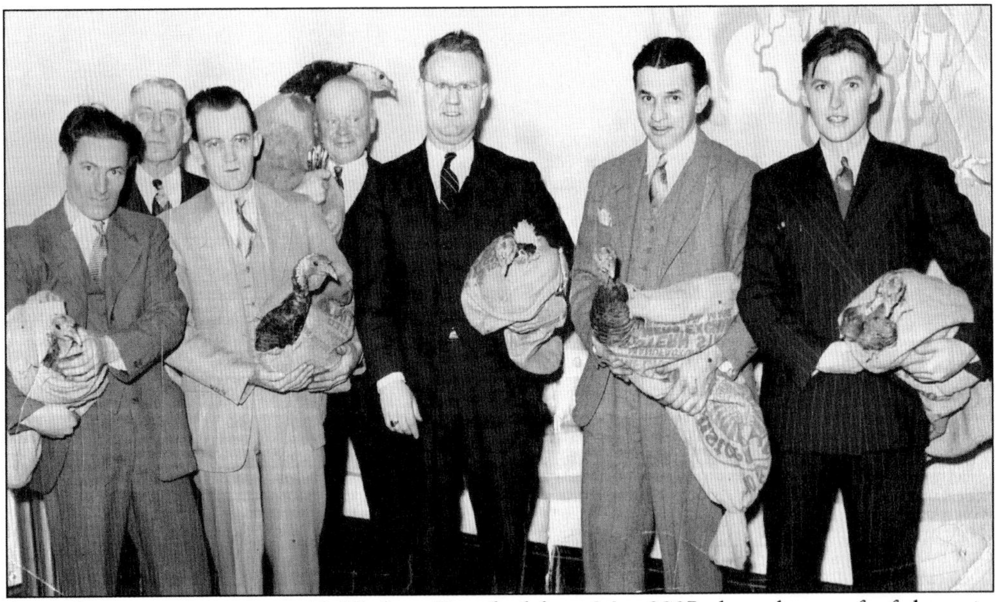

According to an interview with Herbert Crawford from May 2007, the volunteer firefighters in 1938 held fund-raisers to buy equipment. Residents paid to play the card game whist, and the winners won live turkeys. The Graham farm donated the turkeys. In this photograph the men are shown happily holding the prizes. Pictured are, from left to right, (first row) Charles Bunton, David Skelton, Al Cronin, Robert Carpenter, and Everett Graham; (second row) Orray Skelton and Dave Lundin, all volunteer firefighters. (Courtesy of Albert Fay.)

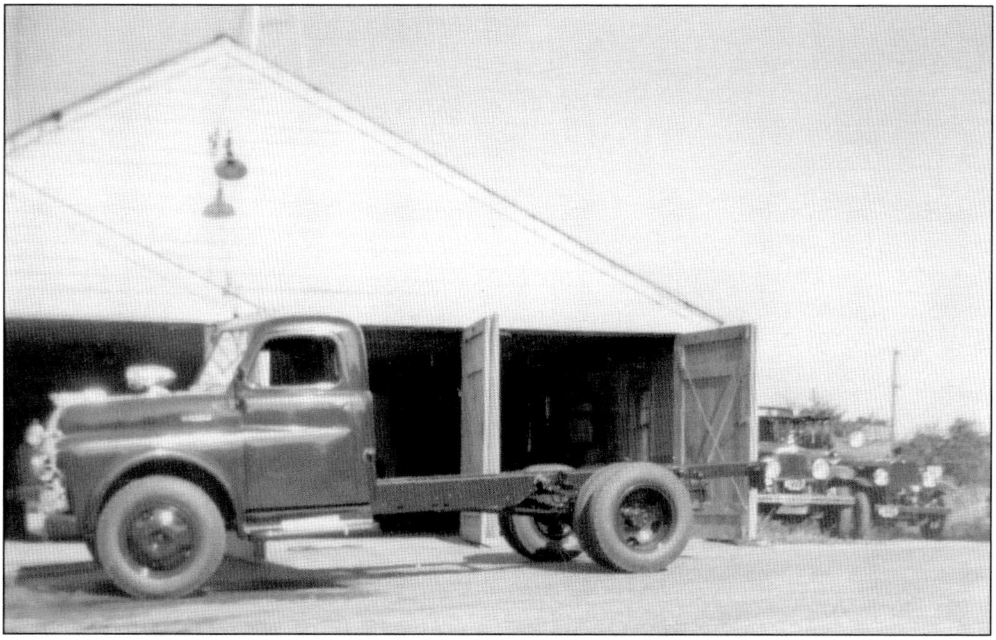

Shown is step one in creating another vehicle, engine 3, by the call department in 1948. These vehicles were put together in the spare time of the volunteers. Stripping down the vehicle and preparing it for the makeover is seen in this photograph. Gaining experience as they progressed, they managed to reach their goal of making new equipment as needed. (Courtesy of Herbert Crawford.)

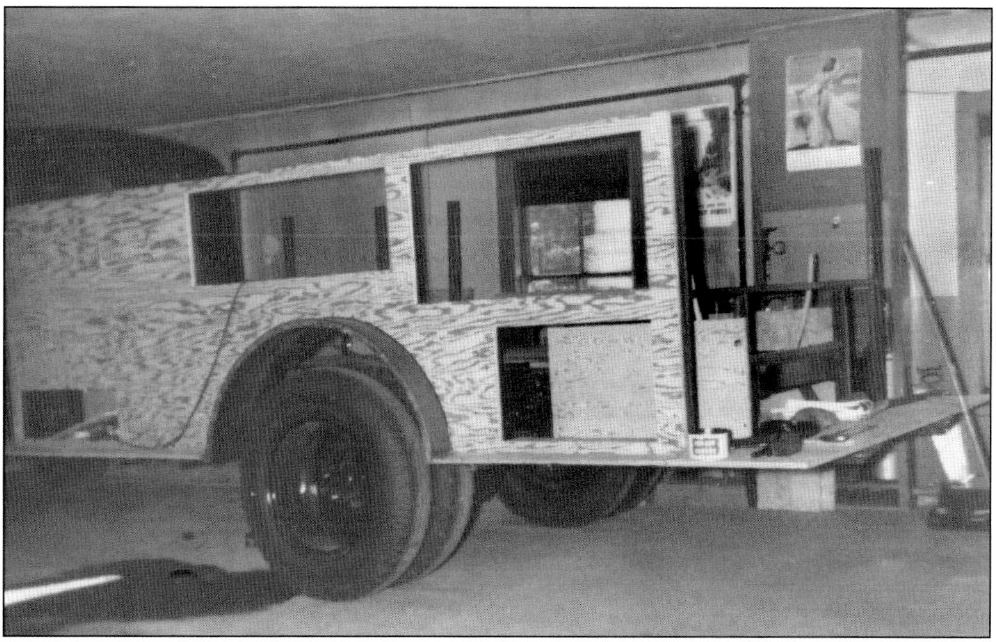

This is step two in the building of the vehicle. Due to shortages in building materials after World War II ended, the men were forced to improvise and use alternate materials. In this photograph, plywood was used in place of metal to complete the back of engine 3. The vehicle was used for many years. (Courtesy of Herbert Crawford.)

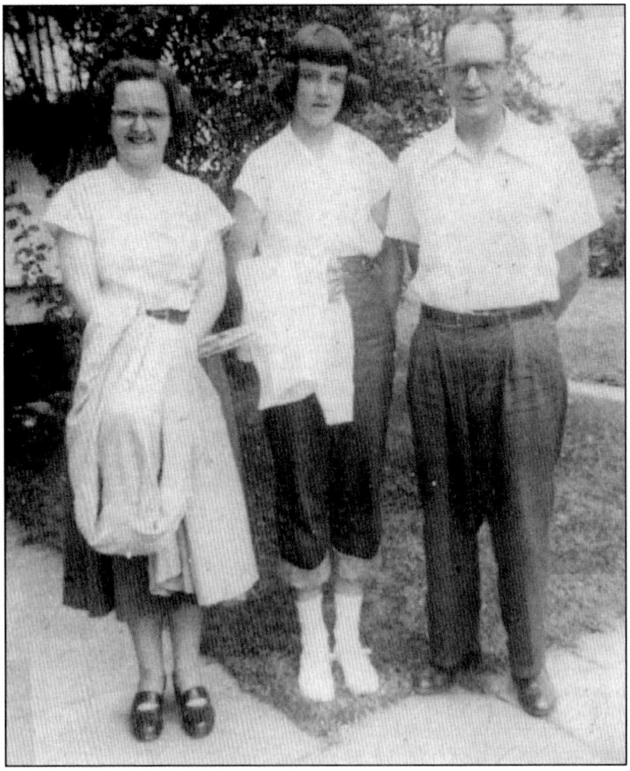

Jane Skelton, her daughter Catherine, and her husband, Fire Chief David Skelton are pictured here. Jane was a licensed radio operator and called firefighters to fires. Jane lived nearby on Sears Street. When a fire was discovered, she rushed to the fire station and began calling the volunteers. Burlington was the first town in Middlesex County to operate a two-way radio. (Courtesy of Burlington archives.)

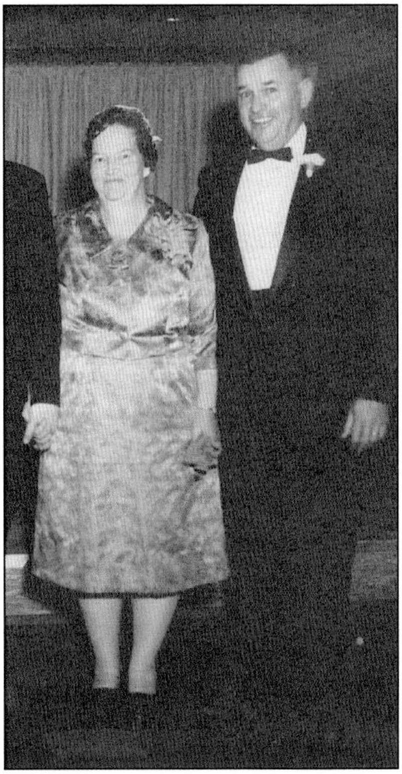

Alice Carpenter is seen with her husband, call firefighter Robert Carpenter. They lived on Dearborn Road directly behind the fire station. Alice became a licensed radio operator and assisted Jane Skelton in contacting firefighters to fires. Alice was the mother of call firefighter Robert Carpenter Jr. and grandmother of Capt. Scott Carpenter of the present Burlington Fire Department and fire dispatcher Elaine Carpenter. (Courtesy of Burlington archives.)

The red, white, and blue ribbon in this photograph is from the Burlington Firemen's Relief Association, June 5, 1949. The association was formed to help the firefighters in times of need. These ribbons were given out at agricultural fairs and local functions. The fund-raising events were well attended and contributed to the Burlington Firemen's Relief Association. (Courtesy of Herbert Crawford.)

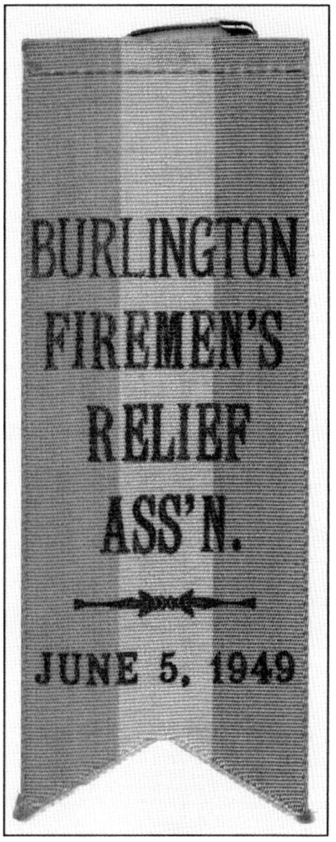

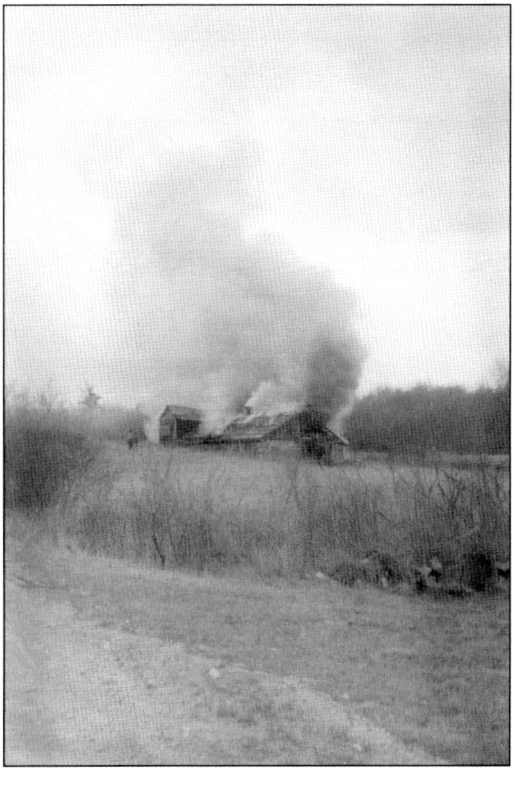

An unusual fire of note was the paint mine fire on Middlesex Turnpike. It completely destroyed the original paint mine on Middlesex Turnpike in the Vine Brook area. The approximate date was 1949. The mine was historically important because Burlington's soil was mined to provide the ingredients for the production of particular colors for the manufacture of paint. It was the only one in Burlington. Neighboring Lexington had many small mines, and today a few remain in its conservation area. (Courtesy of Herbert Crawford.)

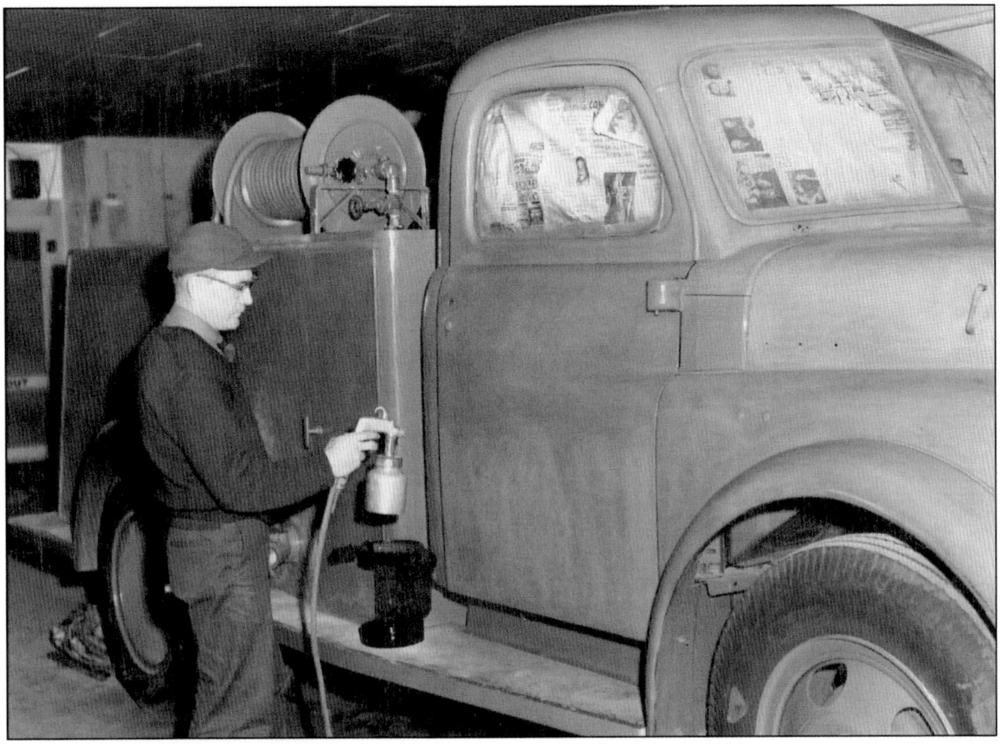

A 1941 Dodge fire truck obtained through army surplus after World War II served as engine 1 until a new 1959 Mack pumper was purchased. The Dodge was relettered as engine 6 until the town bought a Mack pumper in 1962. Several years later, it was painted white as a civil defense truck and was lovingly referred to as the "White Elephant." David Skelton, volunteer firefighter, is shown spray-painting the vehicle. (Courtesy of Albert Fay.)

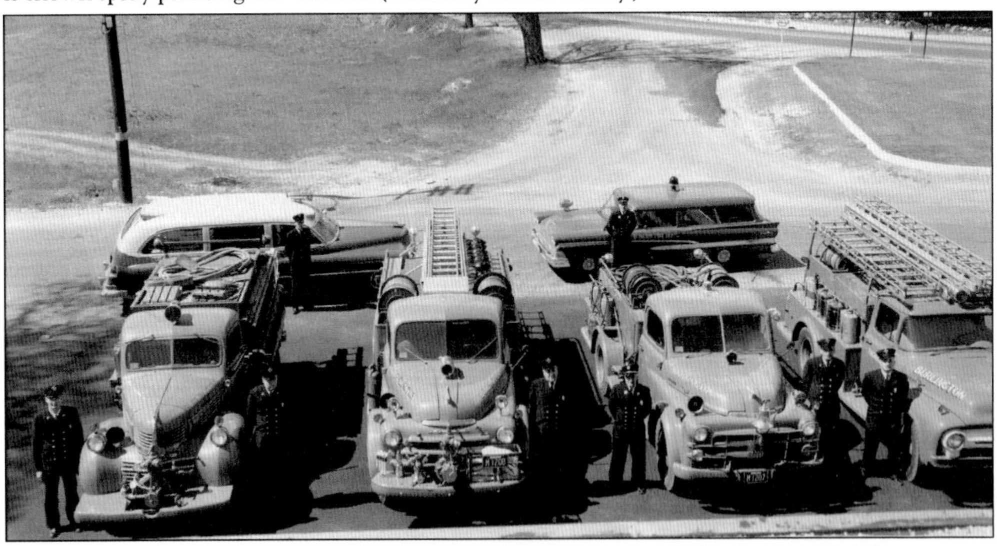

Six vehicles and eight men are on display at headquarters in this 1958 photograph. The men are, from left to right, (first row) two unidentified men, Edward Howard, Charles Bunton, Kenneth Brown, and Frederick Osbourne; (second row) Wilbur McIntire and Chief Herbert Crawford. (Courtesy of Albert Fay.)

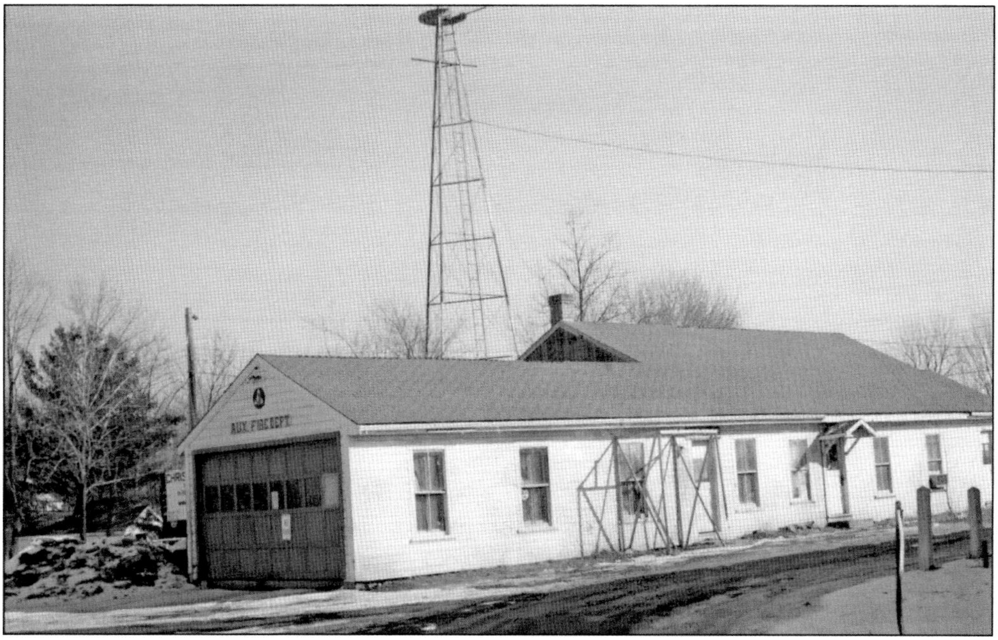

In 1951, a tower at the rear of the old town hall held a horn to sound and identify the location of fires. This allowed call men to respond to the fires by identifying the numbers of horn blasts, which represented sections of town where fires occurred. This made it possible to respond to the location of the fire directly. (Courtesy of Albert Fay.)

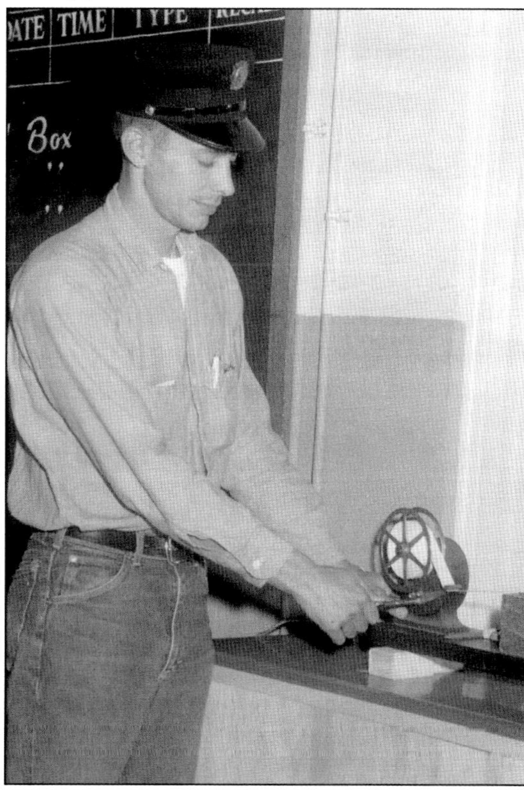

Lawrence Faria is shown with the "tapper" that recorded fire alarm box numbers when an alarm was pulled. The horn or whistle then blew the number of the fire location. (Courtesy of Toni Faria.)

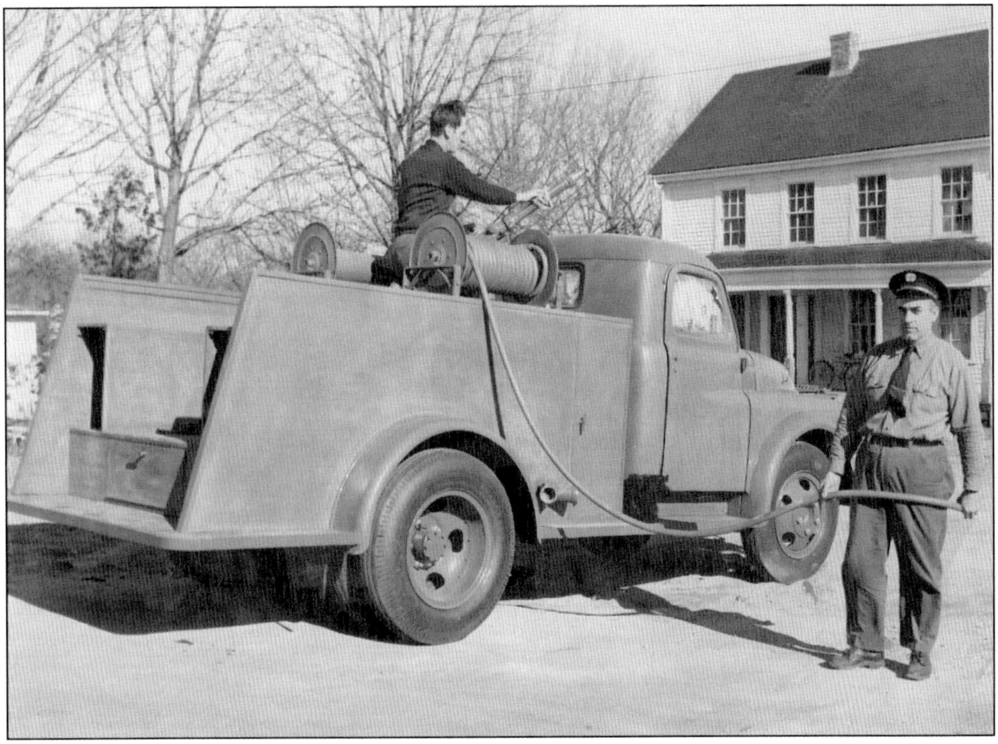

In this 1957 photograph of volunteer firefighters, Wilbur McIntire (left) and Sidney Brown are shown finishing the 1942–1946 Dodge vehicle built by the men of the department. These two men were longtime members of the volunteer department, and Wilbur later became a full-time firefighter. These pieces were later retired to other town departments. (Courtesy of Albert Fay.)

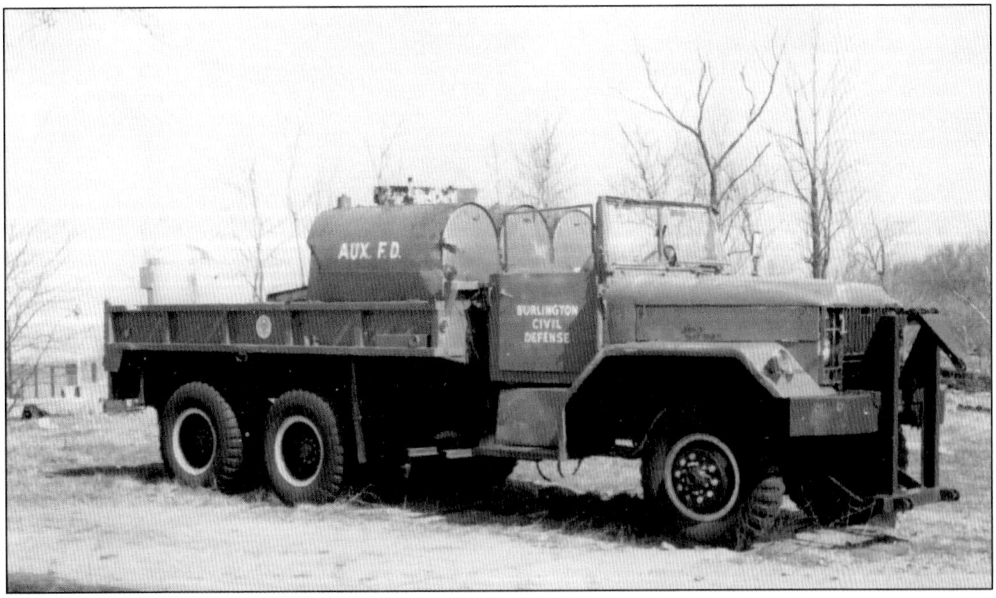

A 1946 tank truck was one of the vehicles that the volunteer department acquired secondhand and managed to turn into effective firefighting equipment. The vehicles were a credit to the ingenuity of the men on the department. (Courtesy of Albert Fay.)

Two

Water District

The threat of losing property to fires continued in Burlington without an adequate water supply. In addition, the need for water was a problem for farmers who usually had enough for their crops, except in dry years when their fields dried up. The constant threat of fires remained, without a water supply. Every farm had its own well, but as the town grew and farms were broken up and the wells and cesspools were too close to one another, it created an unhealthy situation. In 1912, the governor appointed a commission to investigate adequate water supplies in cities and towns. It was the beginning of a long process. It was not until 1932 that Walter Skelton, the local fire warden, spoke at town meeting to authorize the water investigating committee to petition the legislature for permission to "supply itself with water for the extinguishment of fires and for domestic and other purposes." It passed and gave Burlington authority to put in a water system. Unfortunately, because of the stock market crash, no action was taken.

Water holes were dug in Burlington as part of the Works Progress Administration (WPA). The WPA provided programs to help communities in need. Two were completed before the program was terminated. In 1940, another water investigating committee was formed. It was not until 1947 that the Federal Works Agency advanced the funds for plans for the Burlington water system. Work began in 1949. Work began slowly, and many problems hindered the progress. In 1962, the Burlington Water District became the Burlington Water and Sewer Department. During its lifetime, it laid 80 miles of water mains, installed 686 hydrants, connected 4,978 services, operated one main pumping station and seven substations, built two standpipes, and sunk 70 artesian wells. In 1973, a reservoir was constructed on the North Woburn line, and a treatment plant was built to purify the water. The water district lasted 18 years; however, the foundation it put down is a lasting monument to quality engineering and management.

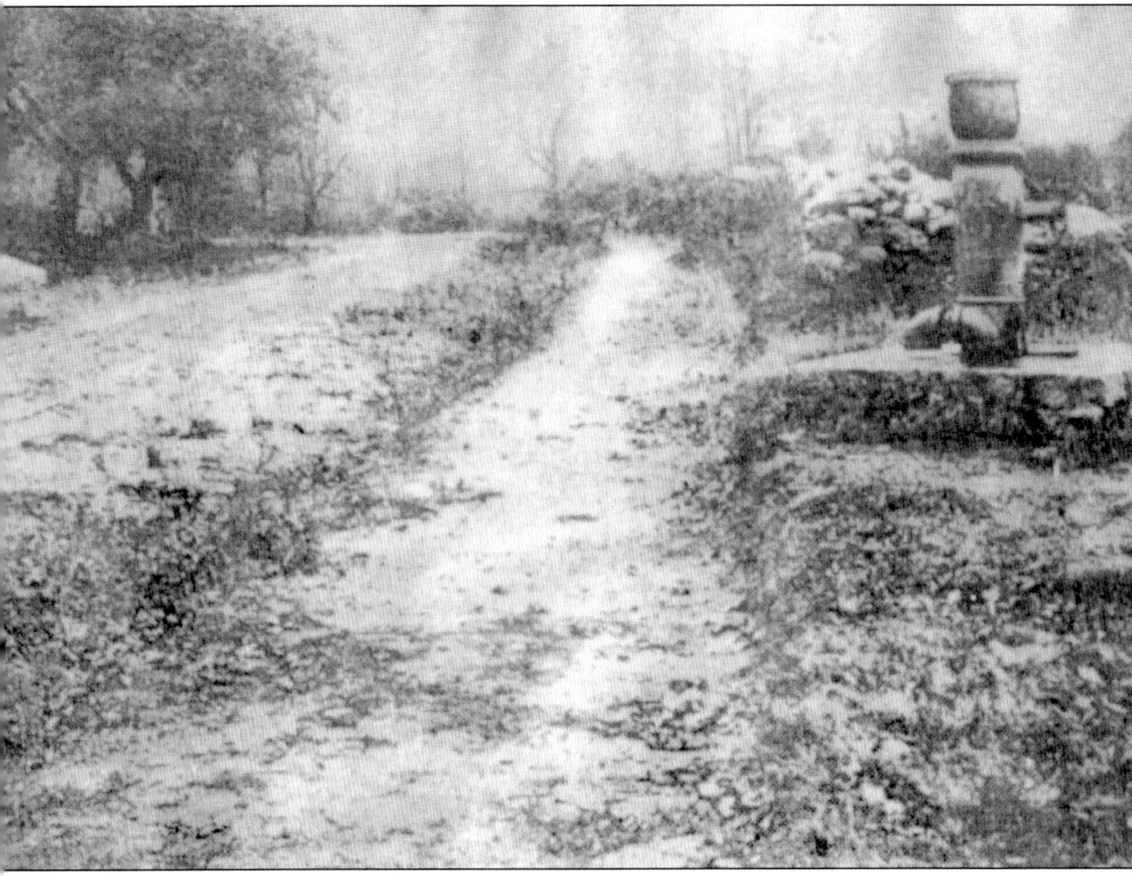

The common held the town pump. This early photograph illustrates the town pump conveniently located in the center of town. An adequate widespread water supply was not provided in town until the water district began in 1949. (Courtesy of Burlington archives.)

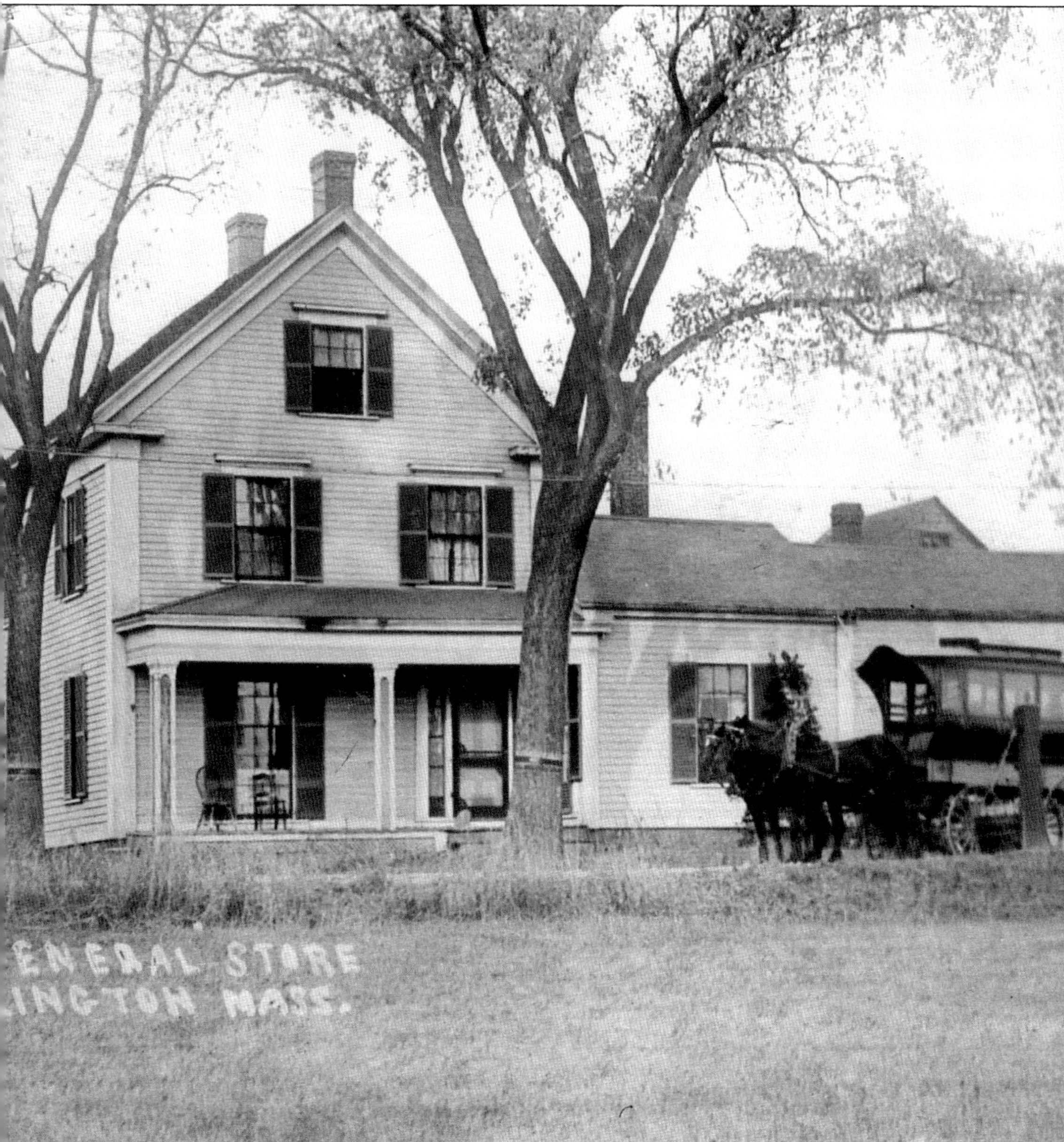

Across from the common was the village store in the Dodge house. An additional pump is in the foreground in front of the town's first school bus, called the "Barge." The capstone was later removed. The lack of water was a problem as the town grew and the need increased. (Courtesy of Burlington archives.)

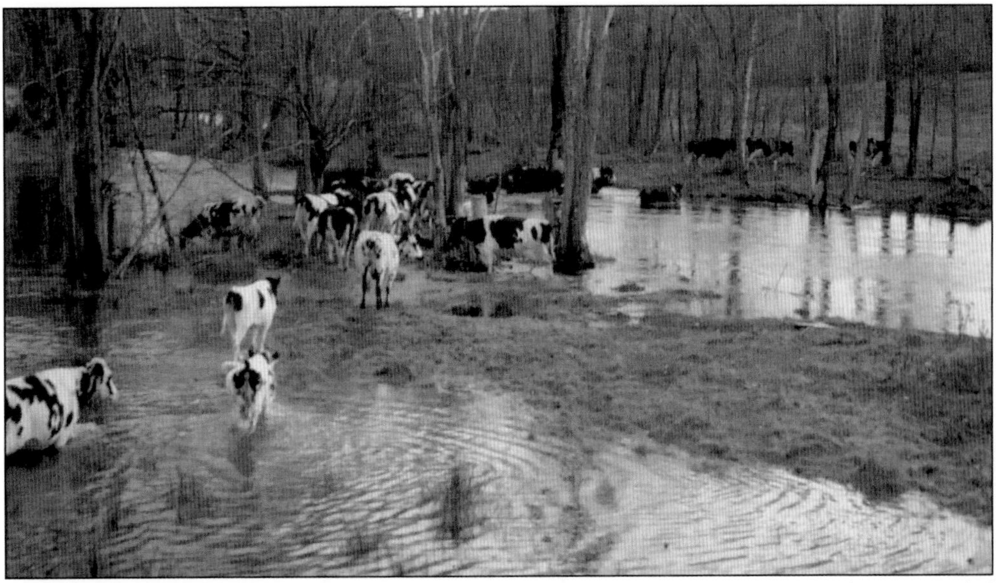

A watering hole for cows on Cambridge Street next to the White construction barn provided water to fight fires if needed. Ponds and watering holes were sources of water to fight fires before the water district brought water under pressure to Burlington. Water holes were later created to provide water in case of a fire. They consisted of a tank buried in the ground filled with water at different locations where needed. In 1939, the WPA built two water holes, one at the Union School and a second one on the corner of Winn Street just before Winnmere Avenue. (Courtesy of Burlington archives.)

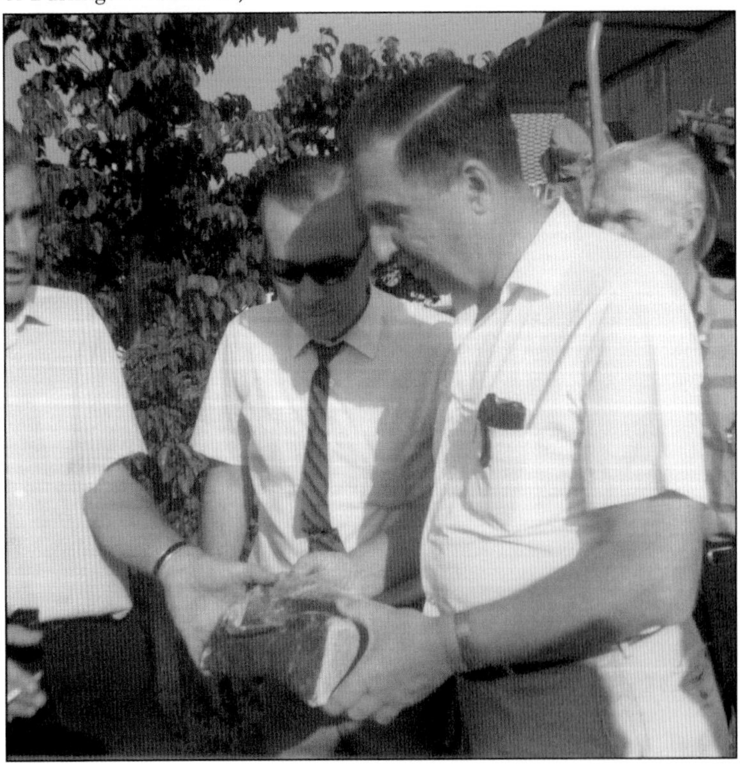

Town treasurer John Fogelberg, shown in this photograph, voted with district clerk Maude Graham, Milton McDonald, Ralph McDonald, and Dave Skelton, water commissioner, to raise $392,000 for funding at the water district meeting in 1949. The leading citizens worked together for a common goal. (Courtesy of Burlington archives.)

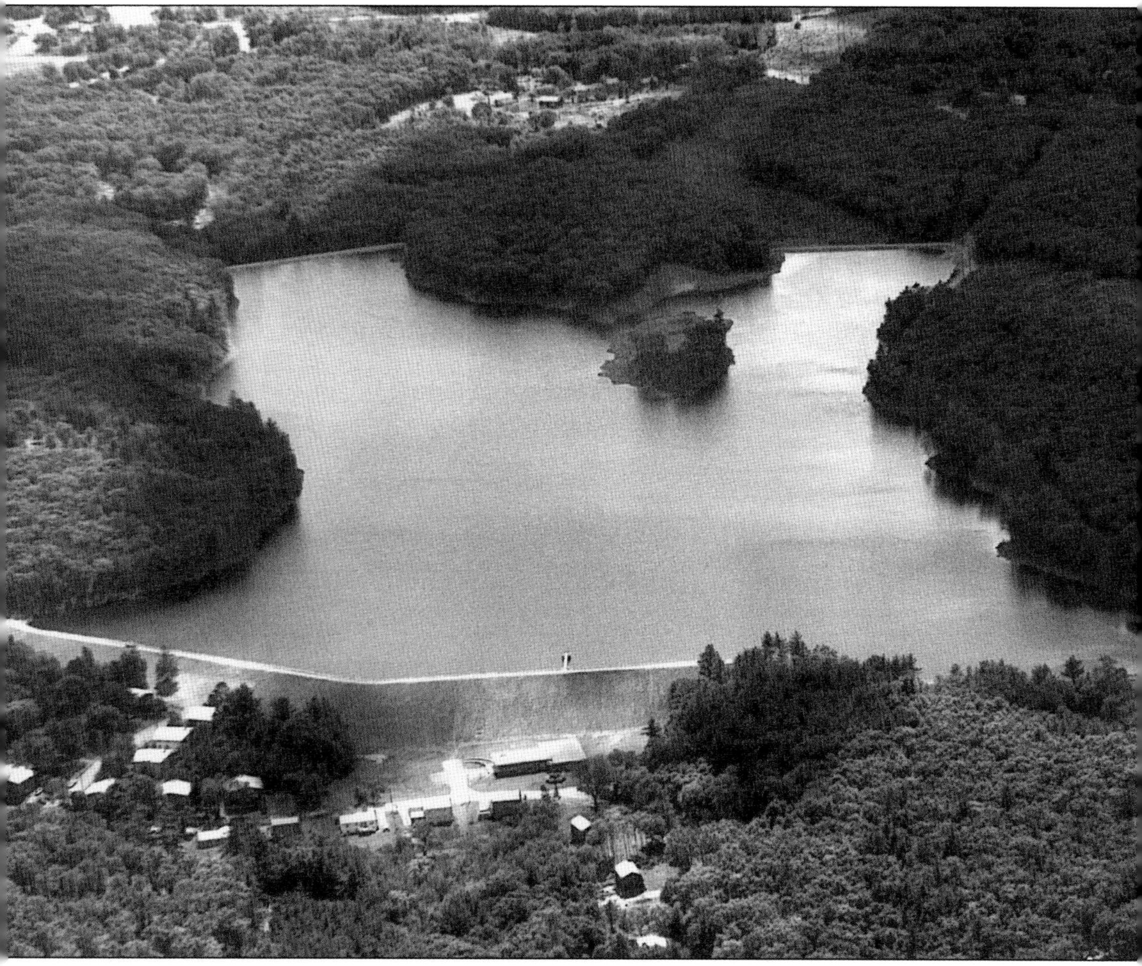

The reservoir that was created as a result of the water district serves the town well by providing an adequate water supply. It was built in 1973. This aerial view of the reservoir shows the size and beauty of the site. (Courtesy of Robert Clougherty.)

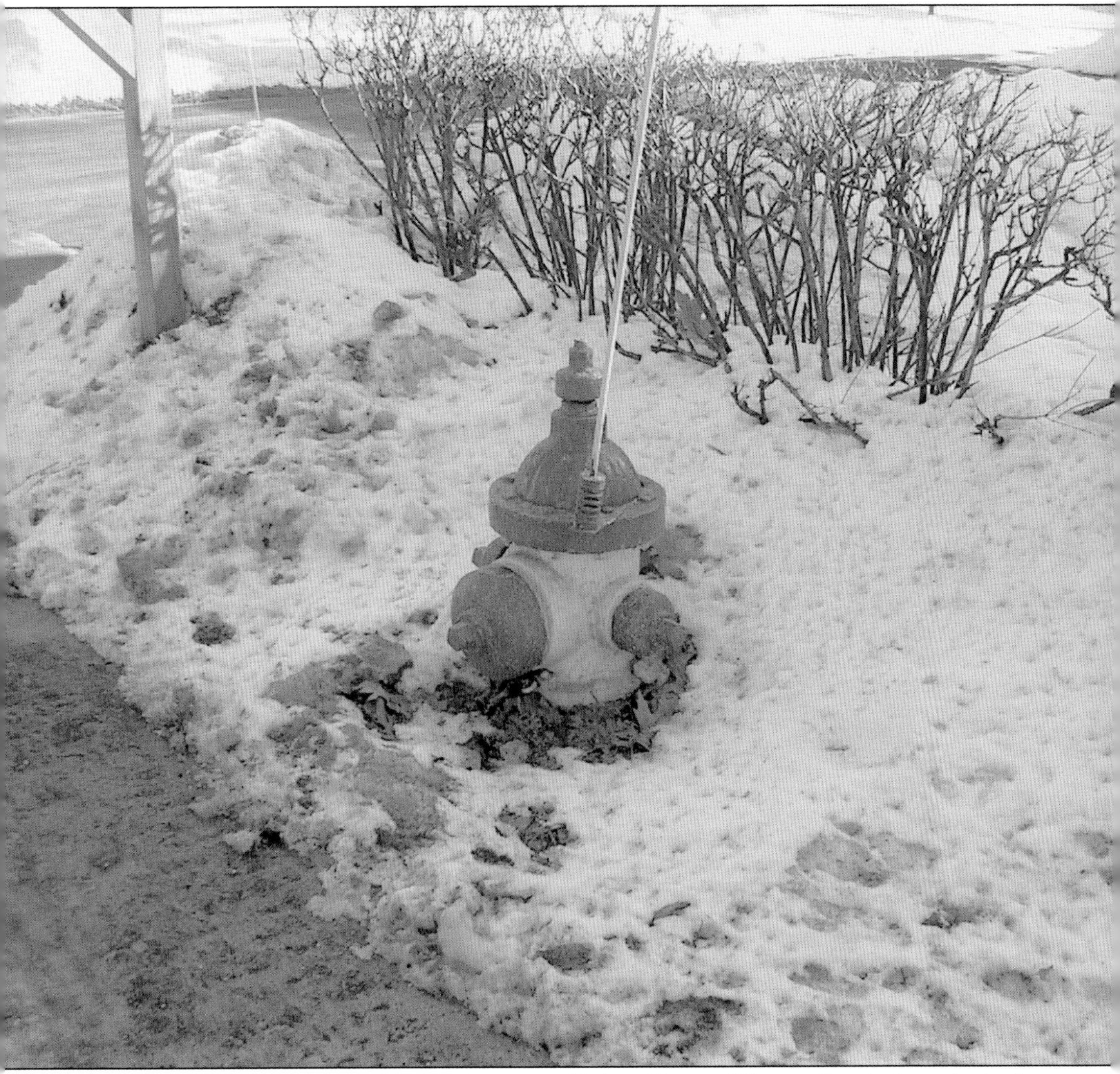

This is a photograph of the last hydrant installed on Francis Wyman Road when the hydrants were installed initially. The water district became the Burlington Water and Sewer Department in 1962 and in 1967 part of the Public Works Department. The installation of 686 hydrants in Burlington brought a feeling of relief to the residents of Burlington who had lost properties to fires due to a lack of available water. (Courtesy of Toni Faria.)

Three

A New Era

Rapid growth in Burlington continued as the town entered a new era. In 1950, an ambulance and resuscitator was bought for $600. Civil service began in 1954. Herbert Crawford was appointed permanent fire chief in 1955. The following year, the force consisted of 7 permanent men and 12 call men. The seven were Kenneth Brown, Charles Bunton, Edward Howard, Wilbur McIntire, Clifford and Louis Skelton, and later Albert Fay. By 1960, the new men appointed were Robert Verrier, Leon Gronquist, Kenneth Mills, Robert Lee, Vincent Pollicelli, Henry Marshall, and Lawrence Faria. In 1962, a new ambulance was purchased. In 1963, 1,019 fire calls were made and 412 ambulance runs were made.

No photographs of major blizzards in 1969 and 1978 remain in the fire department files. Hundreds of motorists were stranded on Route 128. Burlington firefighters performed emergency and rescue services admirably. Off-duty firefighters used their own vehicles to make rescues as well. Firefighter Walter DeCost made rescues with his skimobile. By the 1970s, 26 firefighters received EMT certification. Electronic monitoring equipment and the Jaws of Life were donated to the department. The auxiliary fire department continued under the direction of Steve Duke.

Route 128 came to Burlington and opened Burlington to controlled development. In 1960, the manual force consisted of 15 permanent men and 19 call men. Burlington's economy changed from farming to high technology and retail businesses. High Voltage Corporation, Spray Engineering, and RCA came to Burlington. Between 1950 and 2000, new facilities came to Burlington. The Middlesex Mall, the Burlington Campus of Northeastern University, the Burlington Mall, the Burlington Industrial Center, Northwest Industrial Park, Burlington Cinema, New England Executive Park, Burlington Woods Office Park, Sun Microsystems, and the Lahey Clinic added to Burlington's economic base. Fires and rescues continued, and the need for a modern fire alarm system was evident. Plans for the system were put into motion.

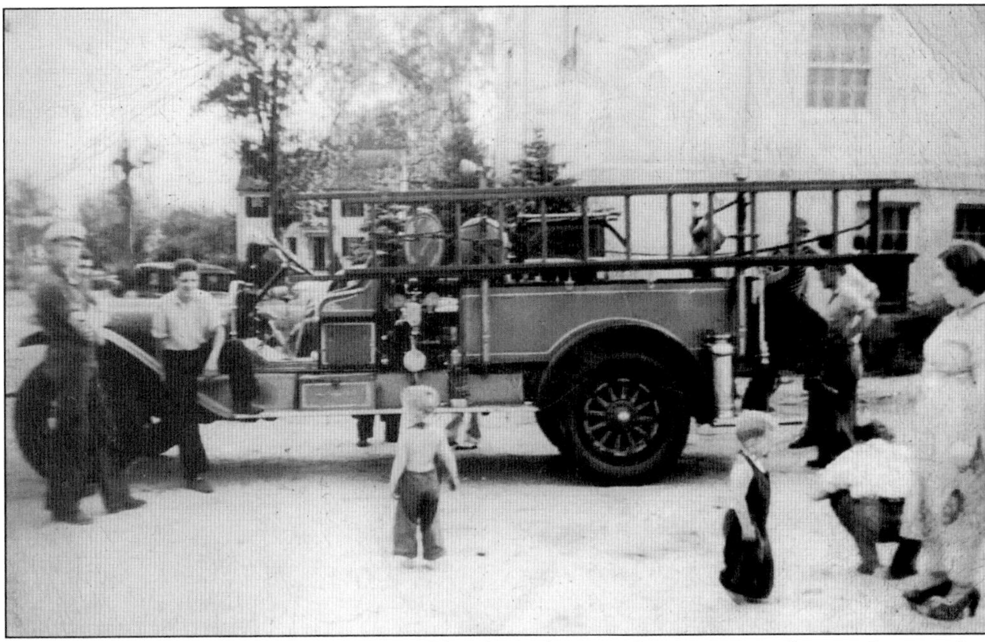

An example of the early equipment used by the volunteer department is this truck, complete with ladders and basic firefighting tools. This is a Charles Bunton photograph. He was a call firefighter in Burlington and later was appointed to the permanent department. (Courtesy of Albert Fay, photograph by Charles Bunton.)

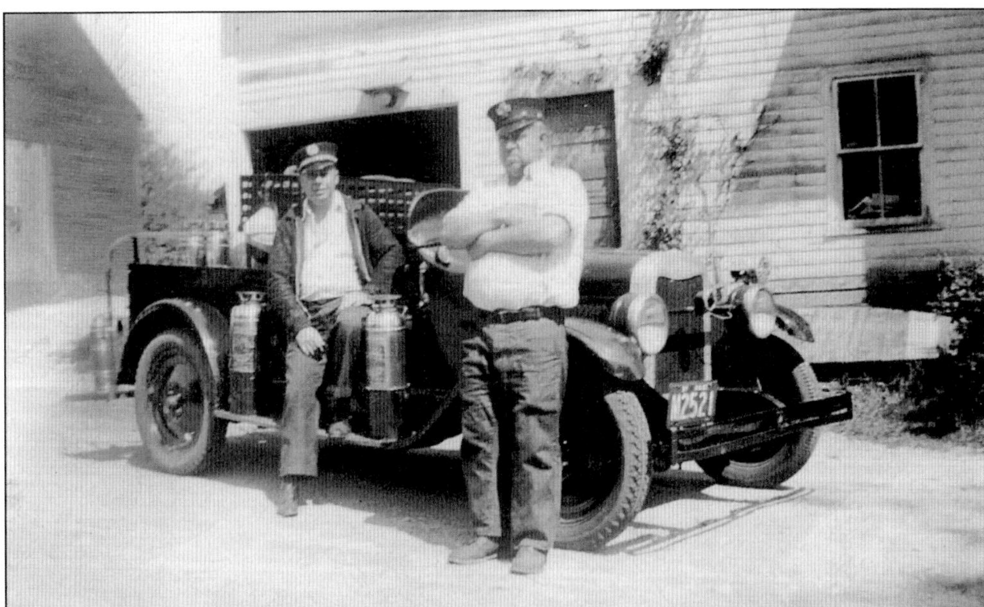

Capt. Theodore Parkhurst (left) and Chief George Martikke pose with this 1935 chemical truck at the Burlington Tire Depot, which was across from what is now the Burlington High School football field. Chief Martikke was the fire chief for one year, from 1937 to 1938. (Courtesy of Albert Fay, photograph by Charles Bunton.)

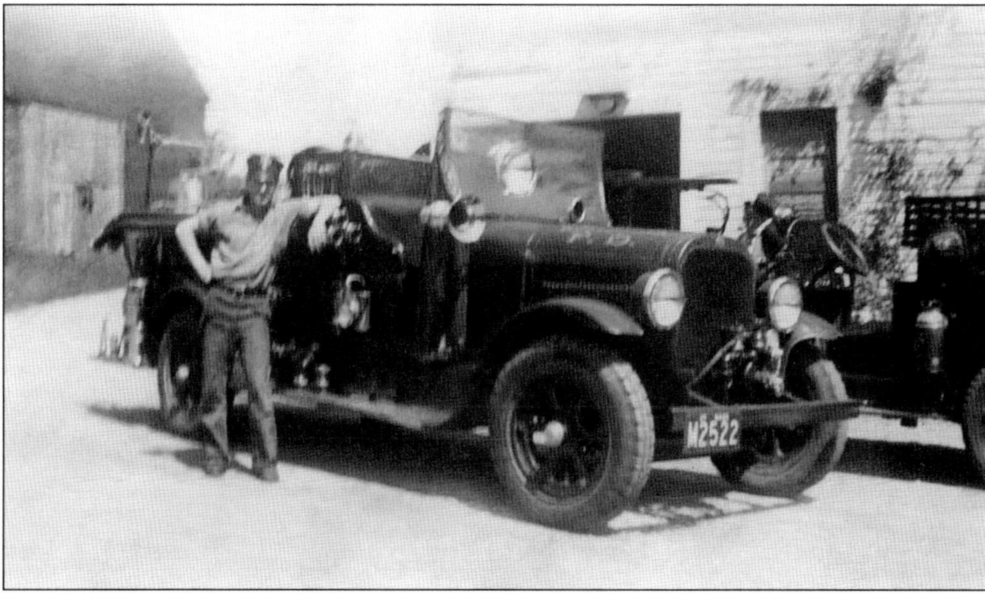

An unidentified volunteer poses with another early vehicle at the same location across from the football field. Charles Bunton, the photographer, was one of the first men to be hired as a full-time firefighter in 1955. (Courtesy of Albert Fay, photograph by Charles Bunton.)

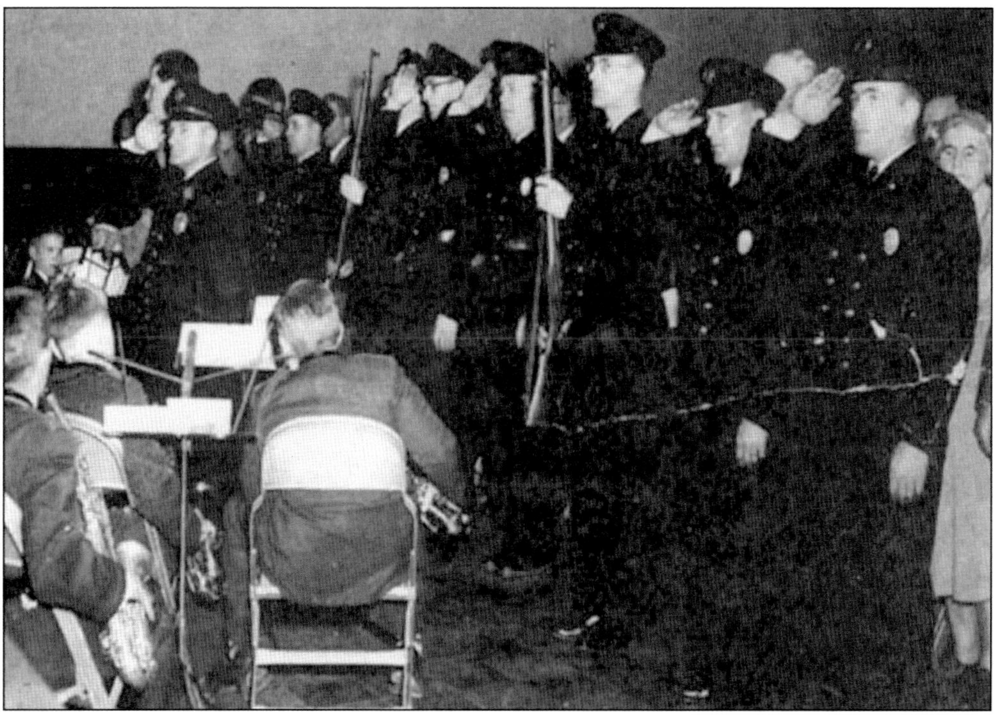

Presenting colors, the fire department donated three flags at the dedication ceremonies of the opening of the Memorial School in 1954. Among those pictured here are teacher Tom Deechan, Deputy Chief Charles Bunton, Chief David Skelton, George Gaudette, rifleman Don Sleeper, Wilbur McIntire, Alan Cronin, Donald Ellsworth, Lester Anderson, Herbert Crawford, and Hilda Oldford. (Courtesy of Herbert Crawford.)

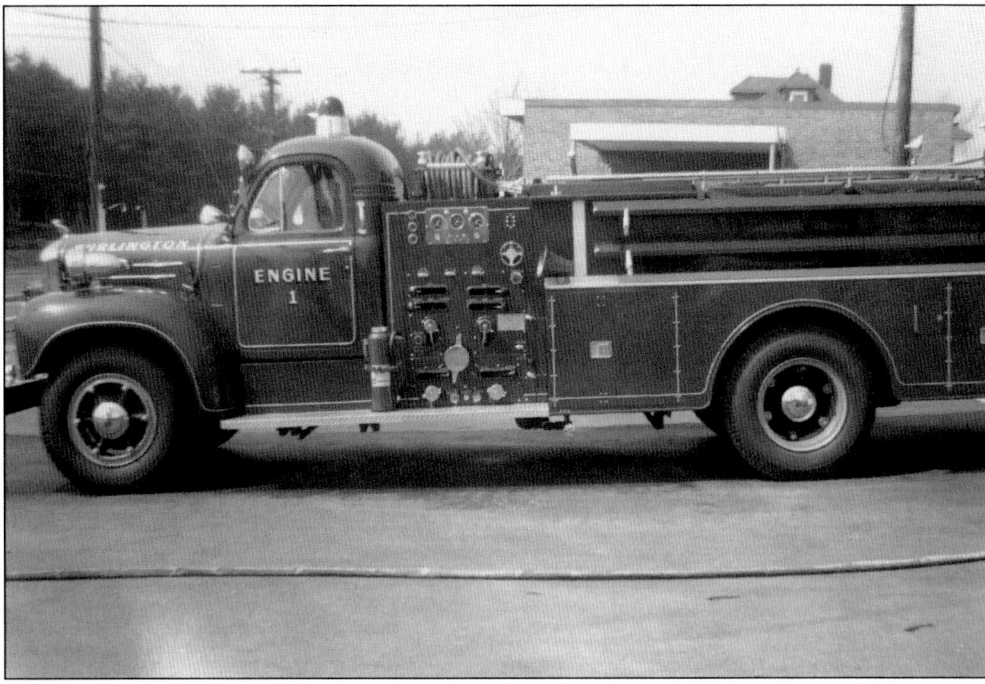

The town purchased engine 1, the first new permanent engine, in 1958, and it was delivered in 1959. The department welcomed the first new equipment, since the volunteers had rebuilt all the previous equipment. The growth of the town continued, and soon more equipment and labor would be needed for the safety and quick response to fires. (Courtesy of Herbert Crawford.)

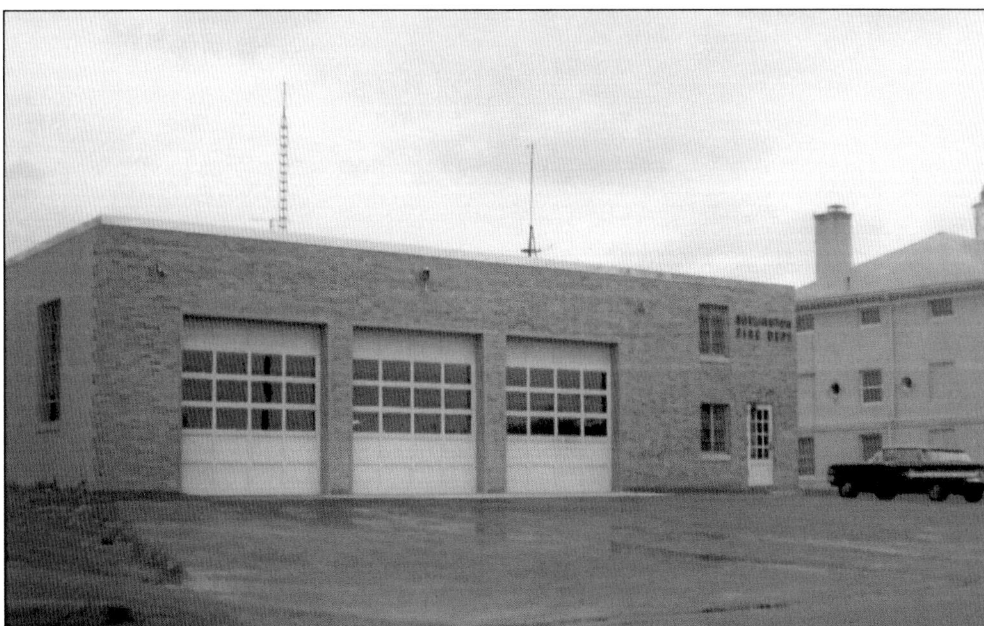

In keeping with the rapid growth of the town, the new fire station opened in 1959 on Cambridge Street across from the common. Engine 1 arrived later that year. The substation on Terrace Hall Avenue followed in February 1971. (Courtesy of Albert Fay.)

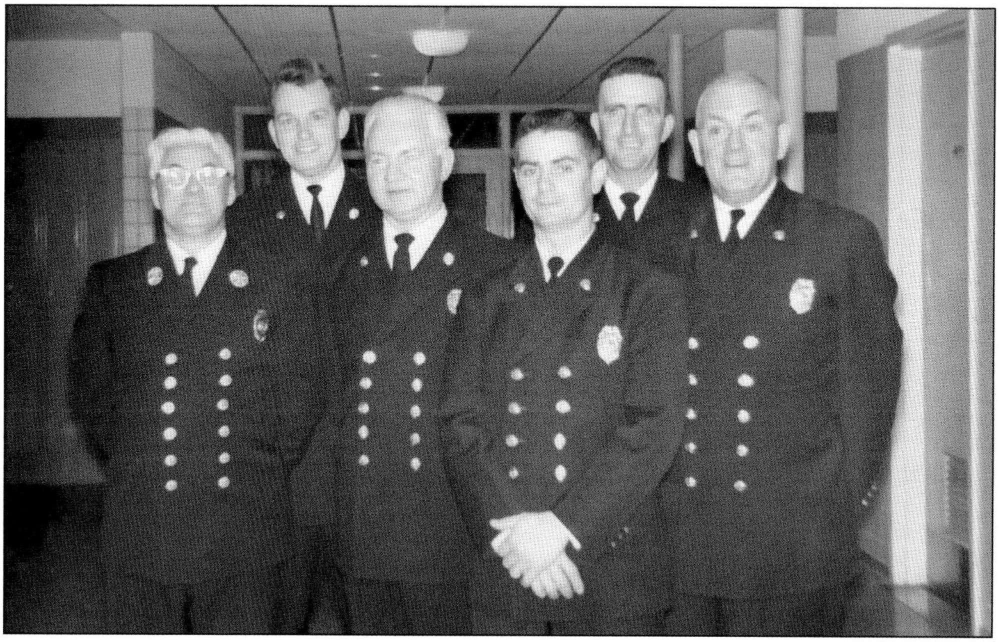

New men were appointed. Three two-man shifts were working a 56-hour week. Shown in this photograph are, from left to right, Charles Bunton, Lester McInnis, Richard Howard, Albert Fay, Kenneth Mills, and Edward Howard. Chief Herbert Crawford worked days with Deputy Chief Charles Bunton. The three two-man shifts were Louis Skelton, Frederick Osbourne, Kenneth Brown, Horace Skelton, Wilbur McIntire, and Edward Howard; later Albert Fay was appointed. (Courtesy of Albert Fay.)

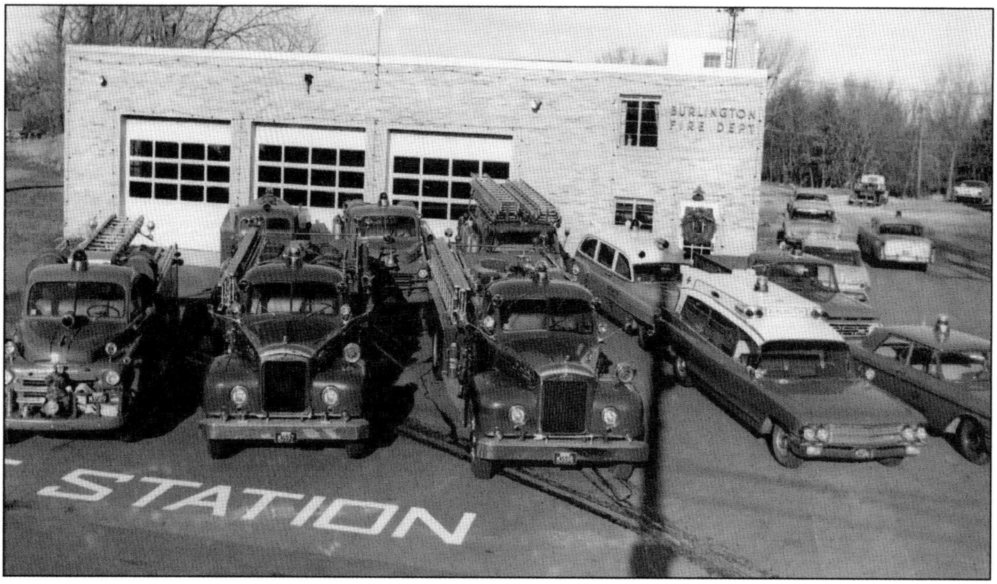

On display in front of the Burlington Fire Department in 1967 are vehicles used in firefighting by the department. Seen here are, from left to right, (first row) engine 3, 1948 Dodge; engine 6, 1961 Mack; engine 6, 1959 Mack; an ambulance, 1962; and the chief's car, 1964 Ford Galaxy; (second row) a tank truck, 1946 Ford; engine 2, 1947 Dodge; engine 4, 1956 Ford; an ambulance, 1952; and a Ford pickup. (Photograph by Al Carbone.)

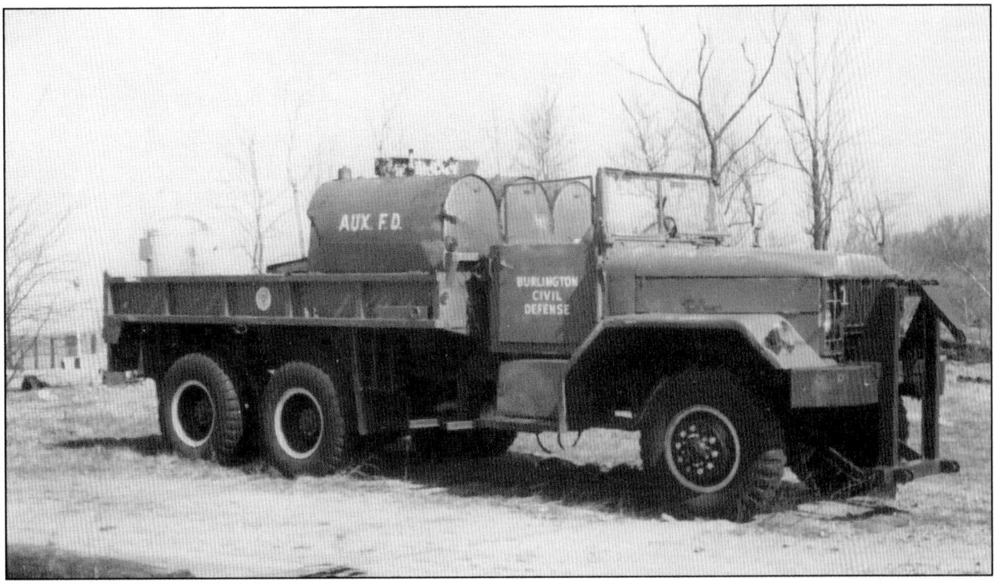

An early fire auxiliary vehicle served the auxiliary for many years. The auxiliary department held meetings in the Dodge house on Center Street, which also served as the civil defense office. Dave Skelton was the director, followed by Kenneth Dobbins. When the town built a new fire station in 1959, the auxiliary relocated to a portion of the old fire station, sharing the space with the police department. It is now under the direction of Steve Duke, who has been a member for 40 years. (Courtesy of Albert Fay.)

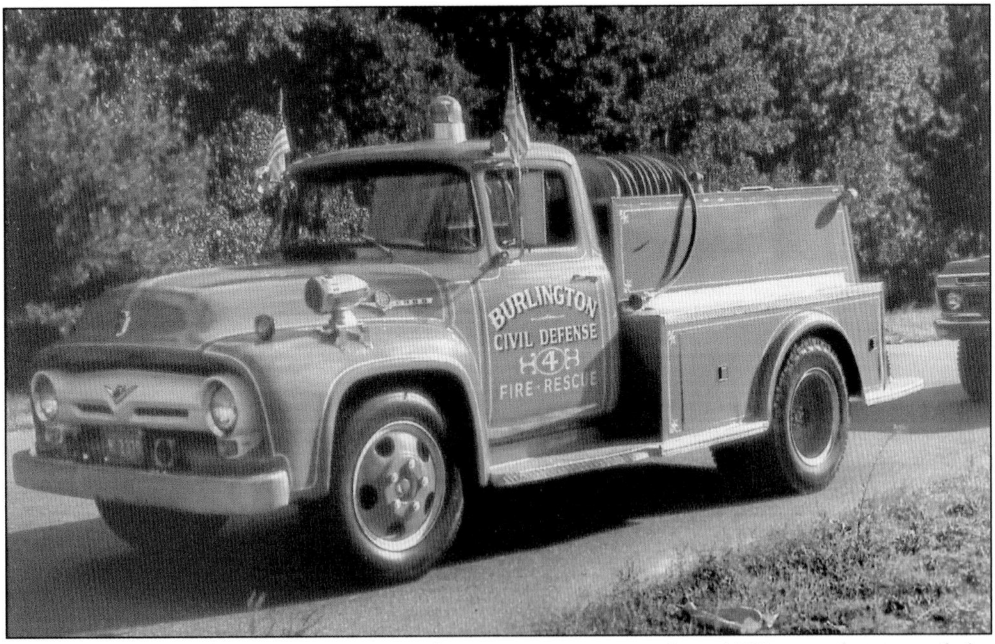

The auxiliary fire department was a volunteer unit organized in 1959 as a division of the Burlington Civil Defense. Today it is the Burlington Emergency Management Agency. A 1951 Ford truck was purchased from Boston Edison for $1, and a generator from the U.S. government surplus property program, also bought for $1, was added. With lights, cables, paint, and lettering, it was placed into service. (Courtesy of Albert Fay.)

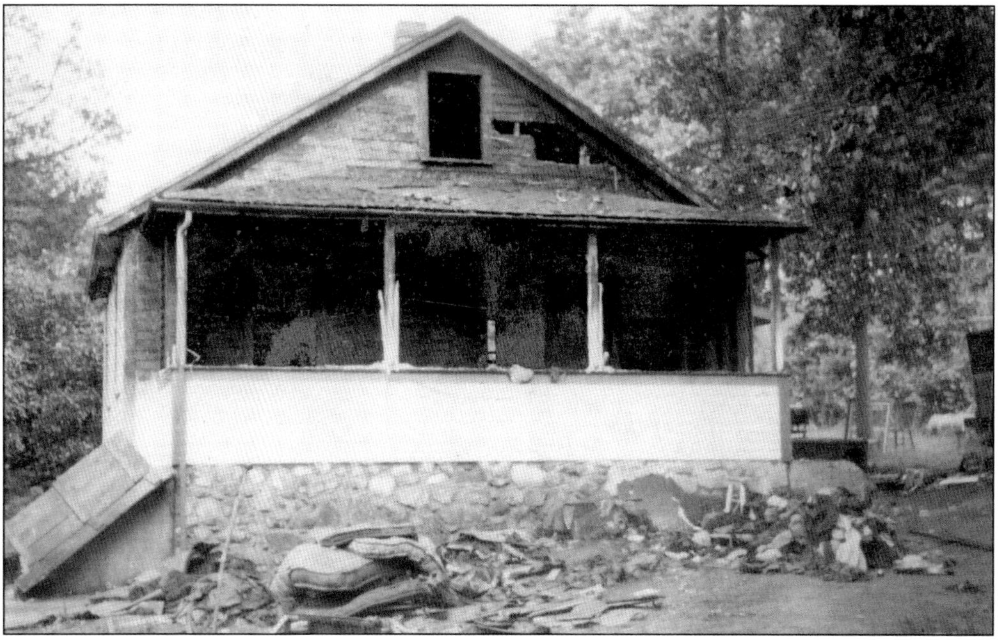

A fire at the home of Vincent Howard on Richfield Street completely destroyed the home in 1954. A gasoline stove burst into flames. The family fled unharmed. The lack of water nearby forced the call men to run a line to Wilmington, the neighboring town, to get water to fight the fire. (Courtesy of Herbert Crawford.)

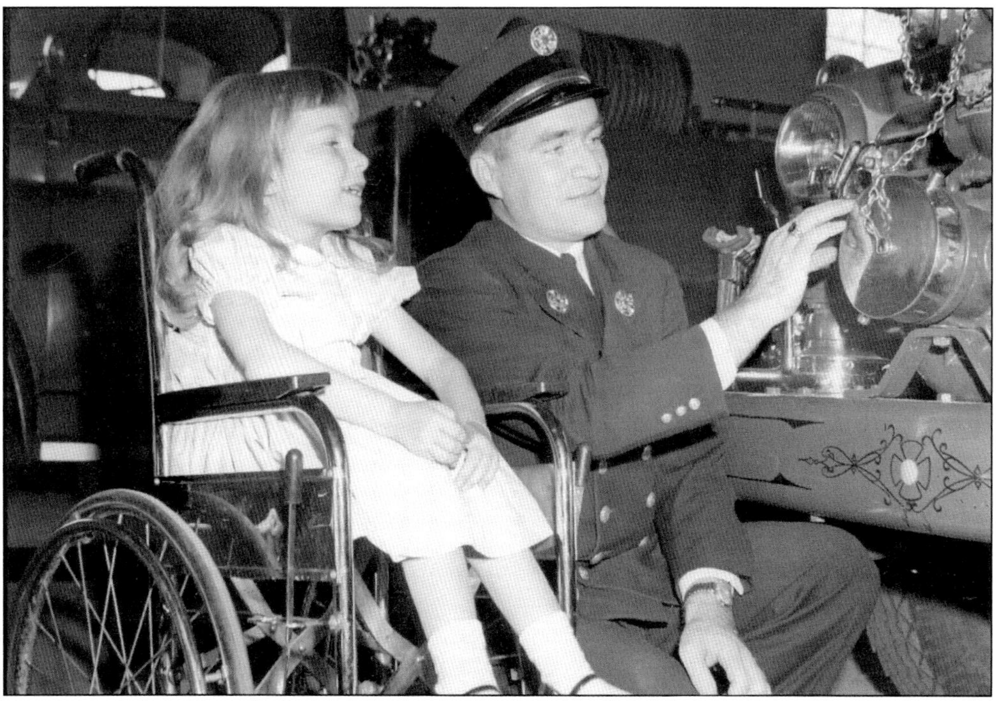

In 1960, Herbert Crawford is pointing out the details of the fire equipment to the muscular dystrophy poster child. Fund-raising for muscular dystrophy is still an annual event in the present-day department. (Courtesy of Herbert Crawford.)

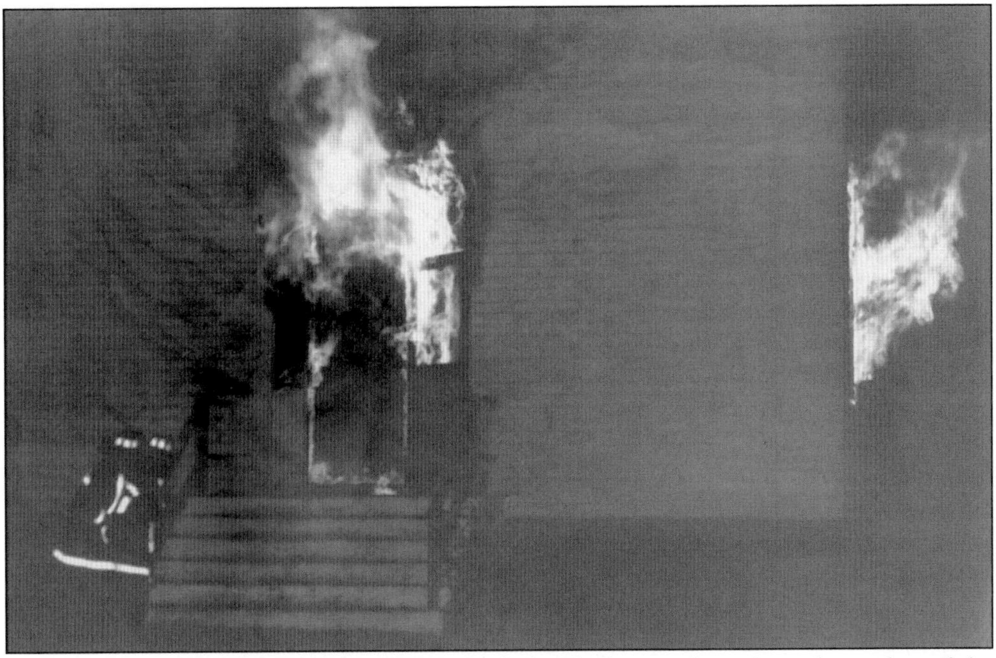

The police station in 1970 was the target of a firebomb thrown through a side window of the building. The building was the old Center School, built in 1855. The police department was housed there temporarily while the new police station was being built. The fire caused extensive damage and destroyed the communications system. (Courtesy of Burlington archives.)

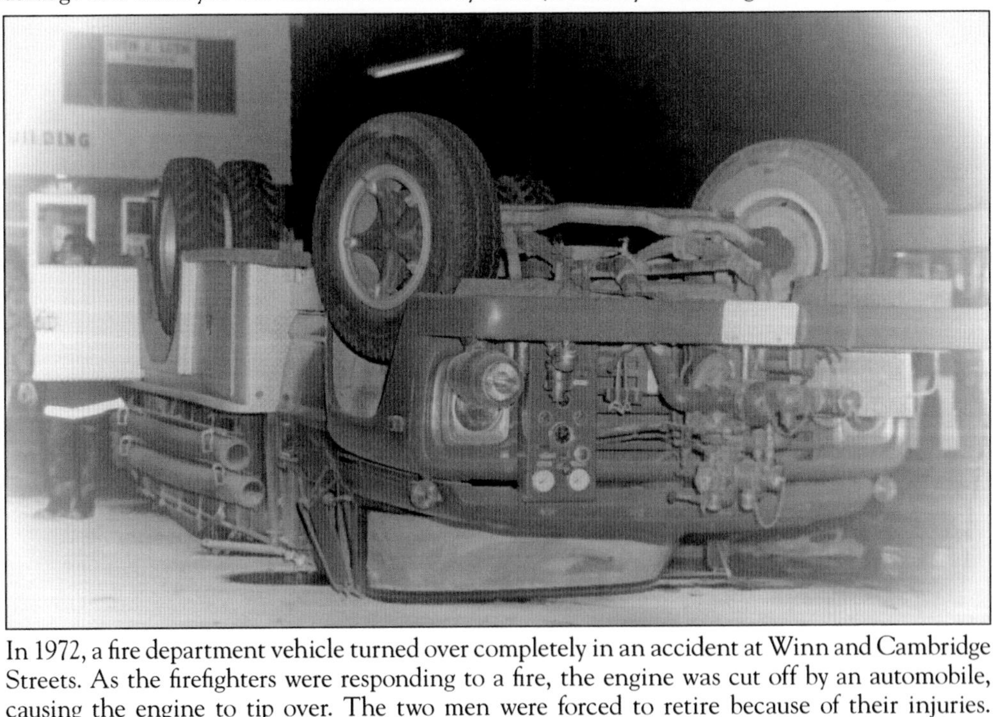

In 1972, a fire department vehicle turned over completely in an accident at Winn and Cambridge Streets. As the firefighters were responding to a fire, the engine was cut off by an automobile, causing the engine to tip over. The two men were forced to retire because of their injuries. Today fire engines must meet stringent safety standards, thus increasing the costs. (Courtesy of Lawrence Faria.)

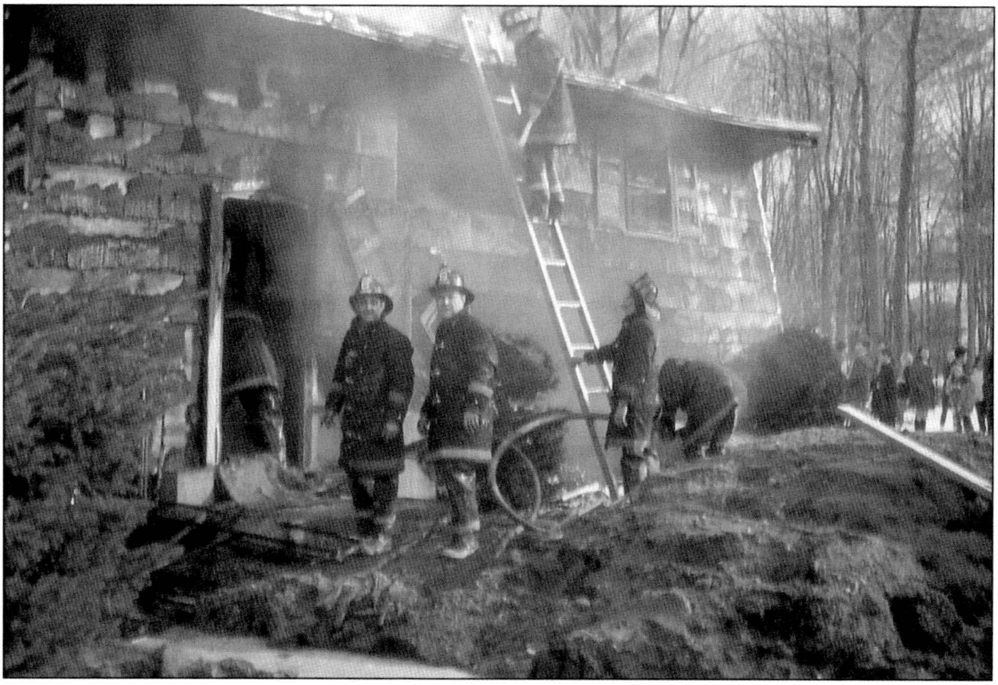

In 1970, the home of the Russo family at Putnam Road was lost to a devastating fire caused by a gas explosion. A contractor digging a trench in front of 1 Putnam Road hit a gas line and ran into the house to alert the residents. The gas explosion totaled the house. The home was fully involved and was a complete loss. (Courtesy of Nellie Russo.)

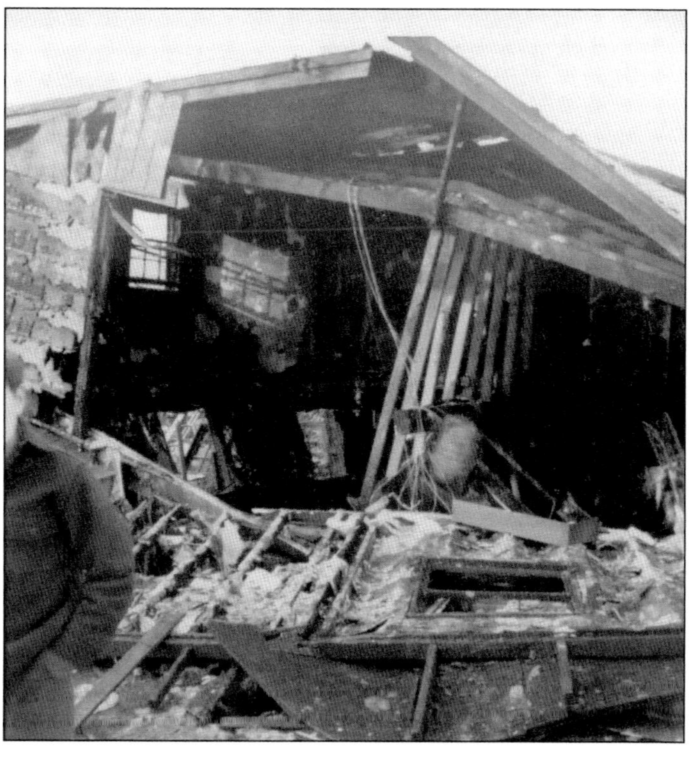

The Putnam Road fire as a result of a gas explosion completely destroyed the house. Fortunately, all the occupants escaped. The damage to the property was extensive, and the home had to be replaced. (Courtesy of Nellie Russo.)

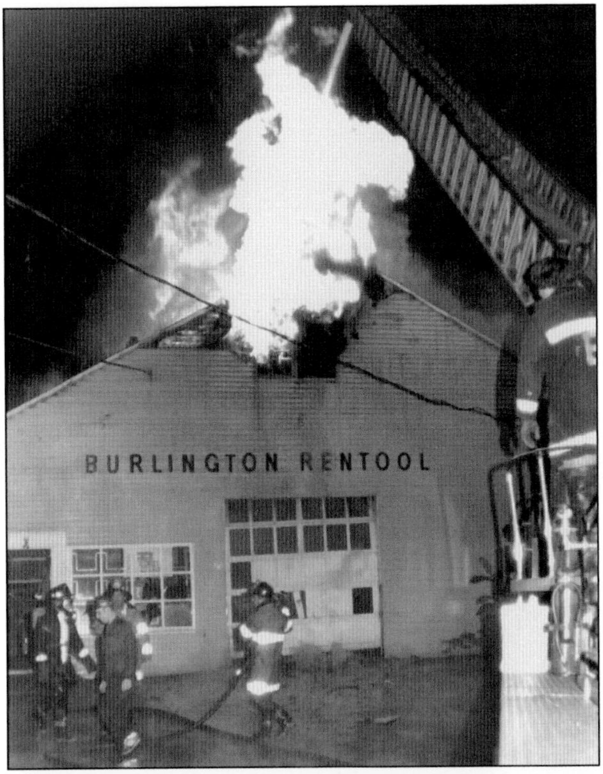

A Cambridge Street fire in 1970 at 128 Electric/Rent A Tool business burned to the ground. It was located in the historic Reed Ham Works building, originally a barn. (Courtesy of Albert Fay.)

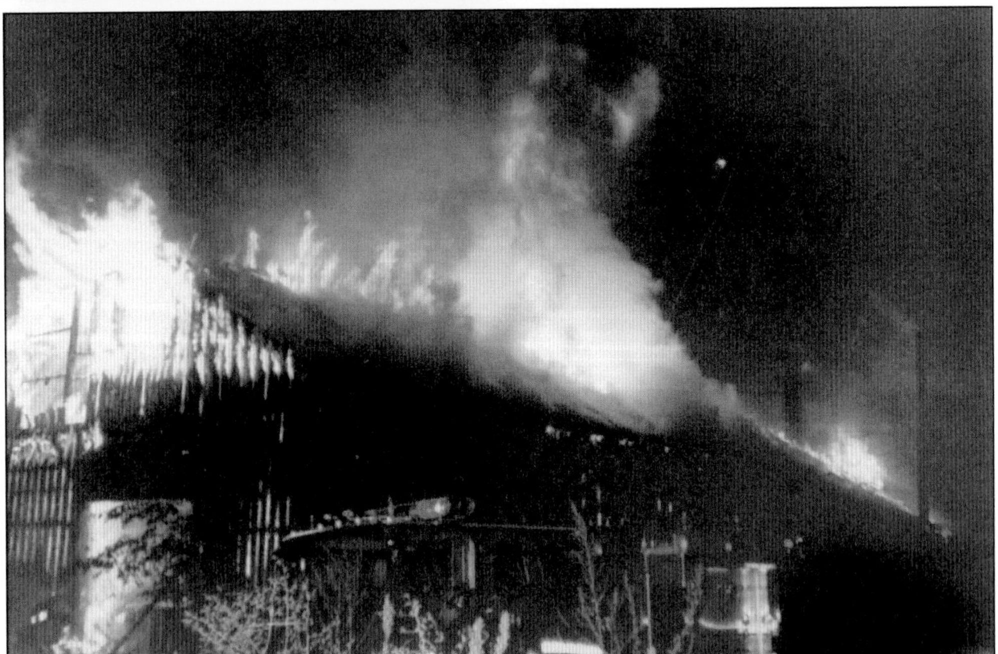

Another image of the intensive fire at the 128 Electric/Rent A Tool building shows the extent of the fire. The fire spread rapidly, and the building could not be saved. The outline of the barn frame can be seen through the flames. (Courtesy of Albert Fay.)

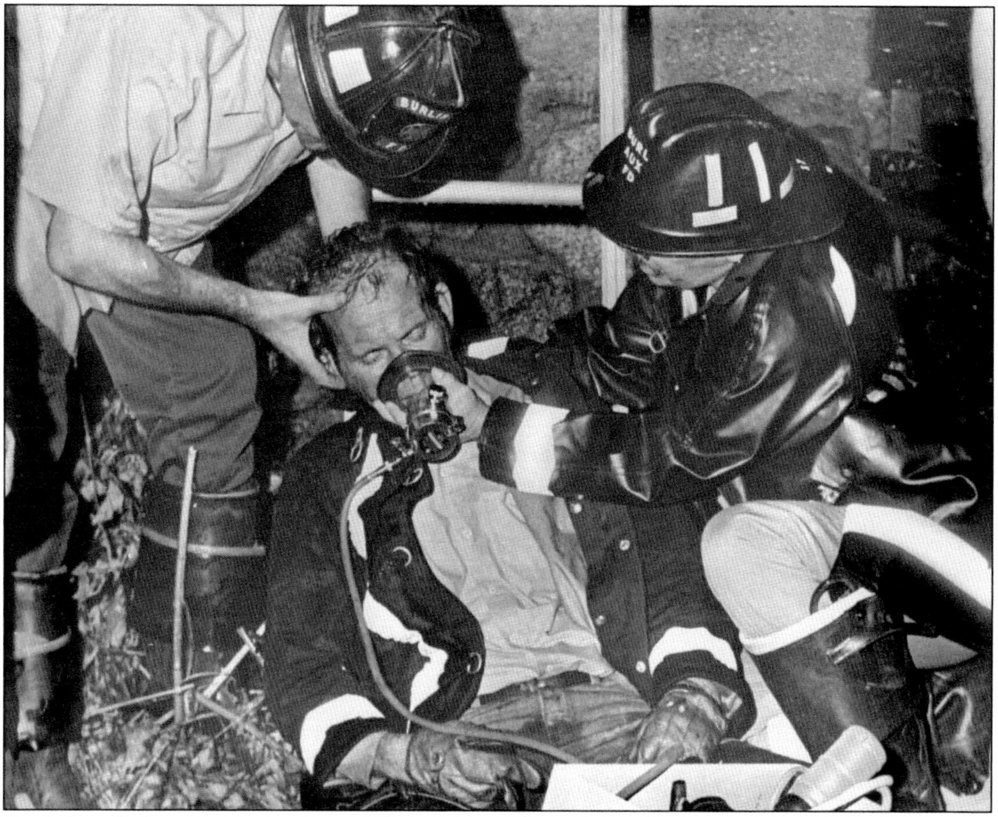

Firefighter Robert Ferro is overcome by smoke at the September 1971 fire at the Hurley property at 1 Winn Street. Firefighter Walter Mountain was also overcome by smoke inhalation. Both men recovered. (Courtesy of Albert Fay, photograph by Al Carbone.)

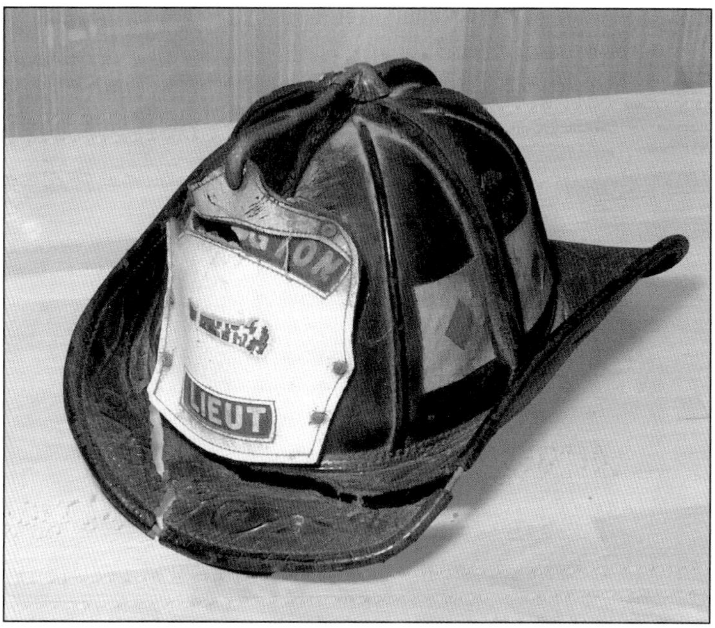

A damaged fire helmet that saved the life of firefighter Lawrence Faria at the Hurley property Winn Street fire in 1971 is shown in this photograph. While on a ladder fighting the fire, he was thrown when a hose and nozzle struck his head. He was injured, but the helmet saved him from further injury. (Courtesy of Michael Tredeau.)

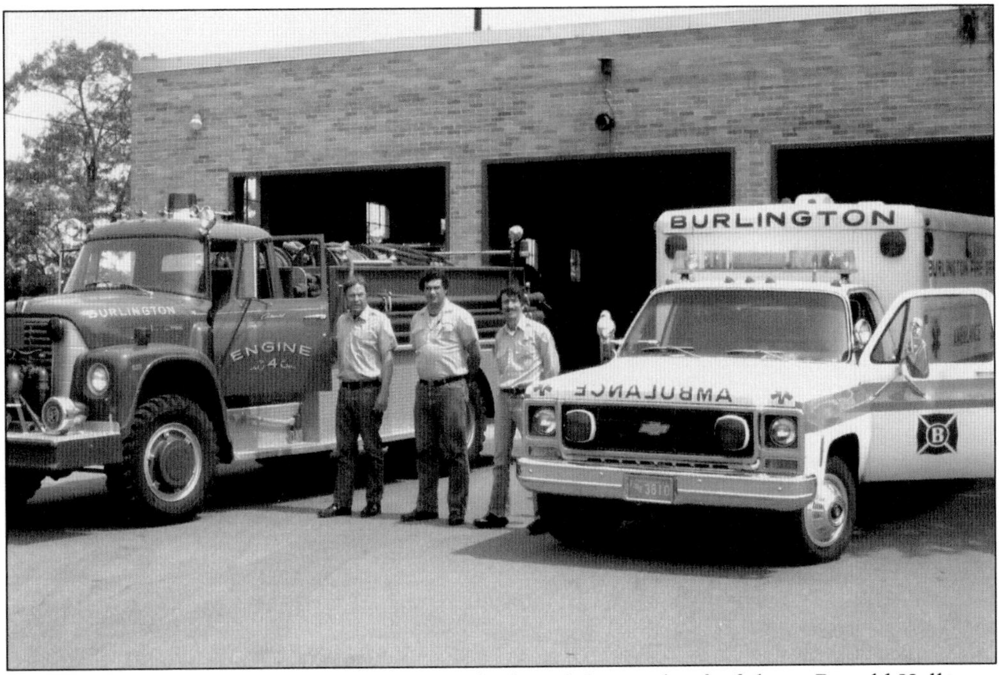

Pictured here is engine 4 at headquarters with, from left to right, firefighters Ronald Kullman, Albert Brabant, and George Creamer in 1981. (Courtesy of Lawrence Faria.)

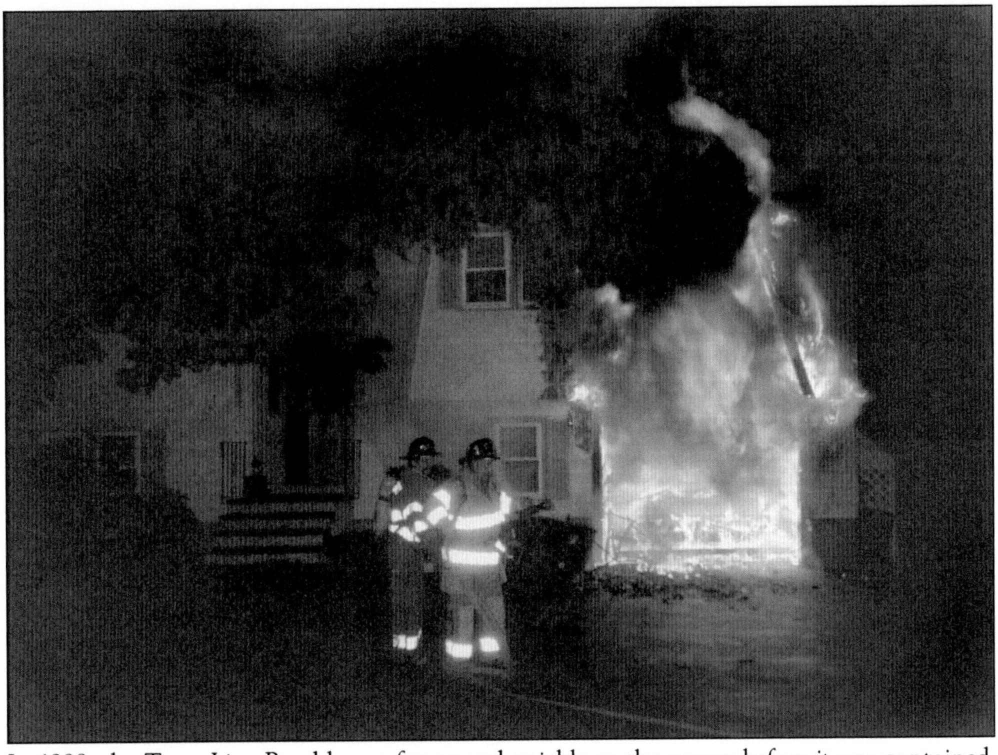

In 1998, the Town Line Road house fire spread quickly to the garage before it was contained. (Courtesy of Burlington Fire Department.)

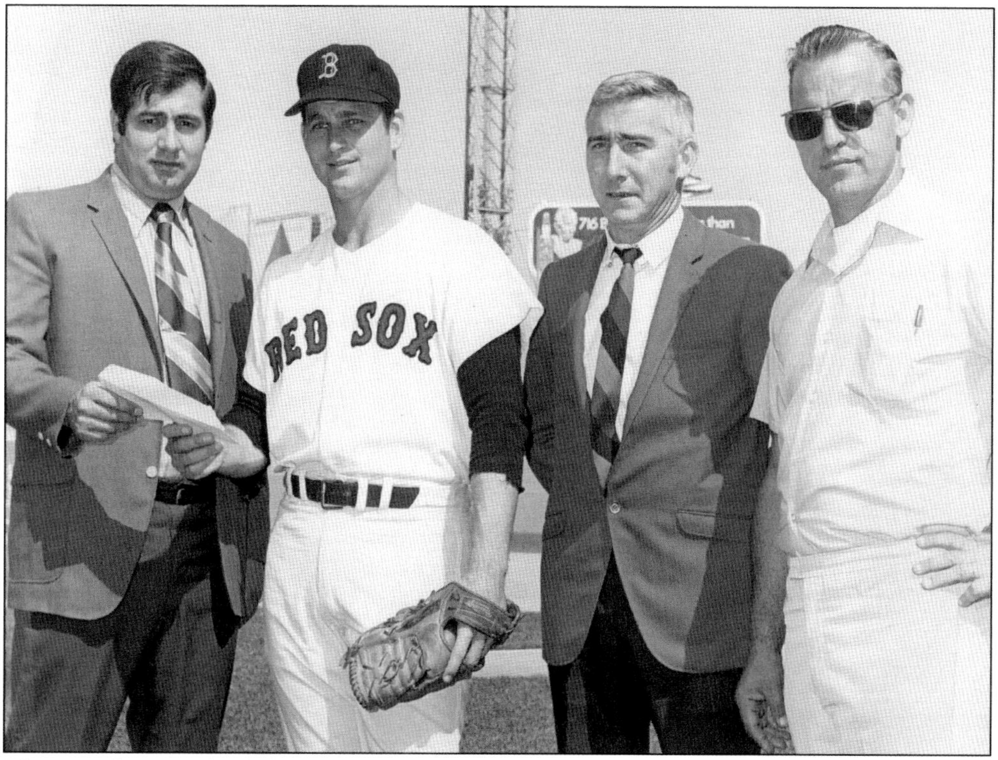

This is a photograph of Phillip Pollicelli handing a check to Ray Culp of the Boston Red Sox for a Jimmy Fund donation from the Burlington Fire Department. Firefighter James Dannaher (second from right) and police officer Carl Christensen look on. Today the department continues to support the charity. The date of the photograph is unknown. However, Ray Culp played for the Red Sox from 1968 to 1973. (Courtesy of Albert Fay.)

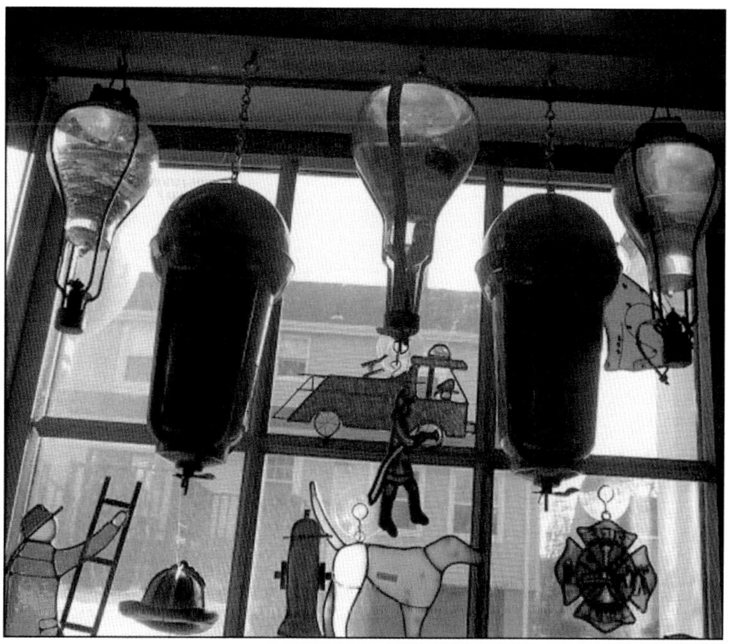

The interior of the Burlington Fire Museum features an assortment of fire-related items. Shown here are antique fire extinguishers made of glass and filled with water or chemicals used in the early 1900s. (Courtesy of Albert Fay.)

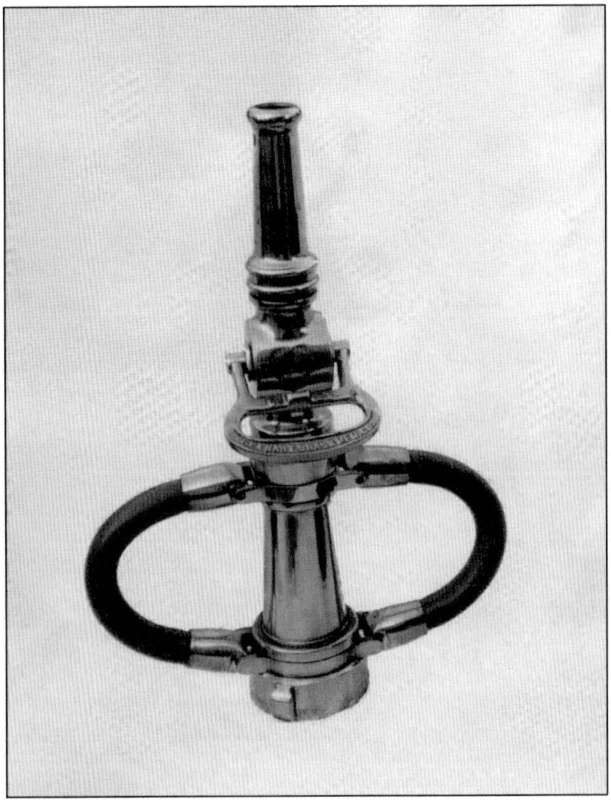

A 1940 brass Play pipe nozzle from engine 3 is shown here. These have now become collectors' items and may be seen in the Burlington Fire Museum. (Courtesy of Albert Fay.)

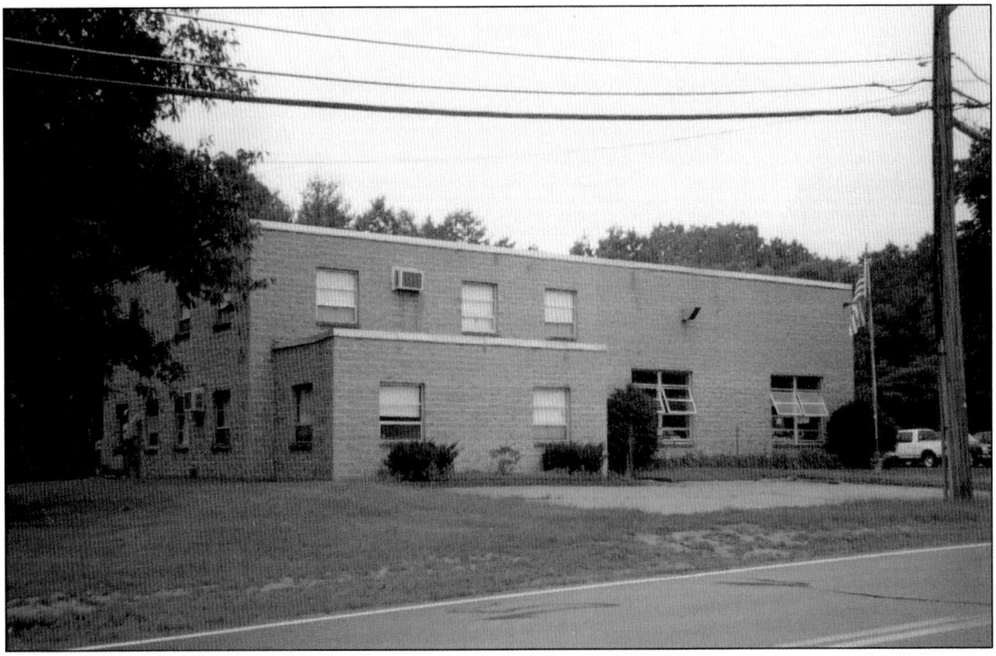

In 1971, the second fire station was built on Middlesex Turnpike and Terrace Hall Avenue. Housed there were new engine 3, a 1971 American La France/Pioneer, a 1,000-gallon-per-minute pumper; engine 4; and a three-quarter-ton four-wheel pickup truck. (Courtesy of Albert Fay.)

Four

Fire Alarm

In 1938, when the dial telephone came to Burlington, it was impossible to reach the firefighters quickly in case of an emergency. Previously telephone operators contacted them. The solution was a fire alarm system. Money was appropriated for a fire alarm transmitter, and in 1939, it was installed.

Later in 1967, when the town established the first fire alarm division, the need for a fire alarm system at the main station was evident. The communication equipment was bought by the town, and some materials were donated. However, it was designed, built, and installed by fire alarm superintendent Joseph Butler.

Together with fire alarm inspector Henry Marshall, he also built consoles, which handled traffic light control, radio communications, EMT alarm systems, and a municipal burglar alarm system. They laid out the circuitry and wiring that connected the station to alarm boxes in banks, shopping centers, municipal buildings, and schools. They also built the consoles in the watch room. It handled interdepartmental communications, traffic light control, radio communications, EMT alarm systems, and municipal burglar alarm systems. The Town of Burlington saved in excess of $500,000 by having the supplies donated and the work done by the two firefighters, Joseph Butler and Henry Marshall. By 1974, the fire alarm division had built a modern communications console.

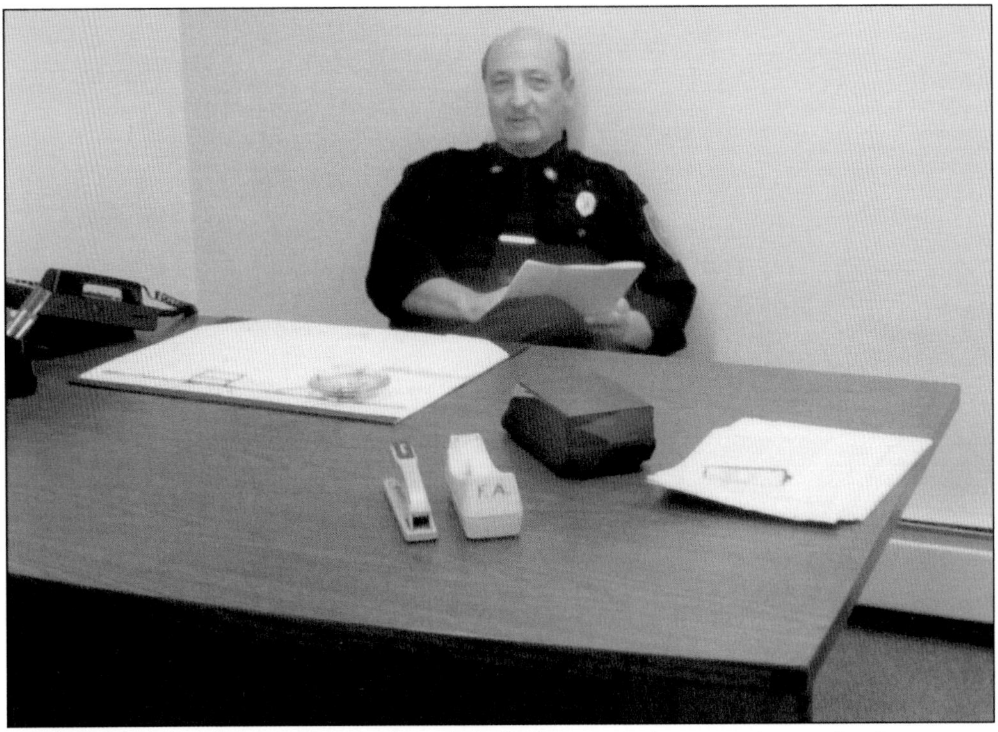

Joseph Butler, fire alarm superintendent, brought his particular expertise to the department, helping to improve and update the communication system. In addition, he was a dedicated firefighter. (Courtesy of Margaret Butler.)

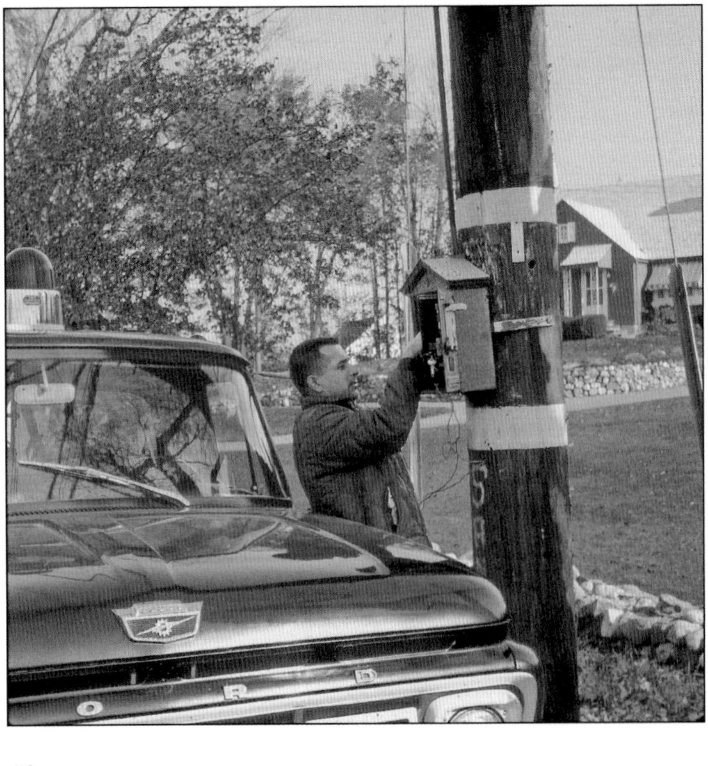

Firefighter Vincent Pollicelli is shown checking a fire alarm box. The new fire alarm system initiated by Joseph Butler and Henry Marshall began in 1967. (Courtesy of Lawrence Faria.)

Pictured here is an antique fire alarm box. Fire alarm boxes operate on specific current—100 milliamps DC. They are wired in a series like many Christmas tree lights, so that when one box breaks the circuit by being pulled for an alarm, the entire circuit opens. (Courtesy of Albert Fay.)

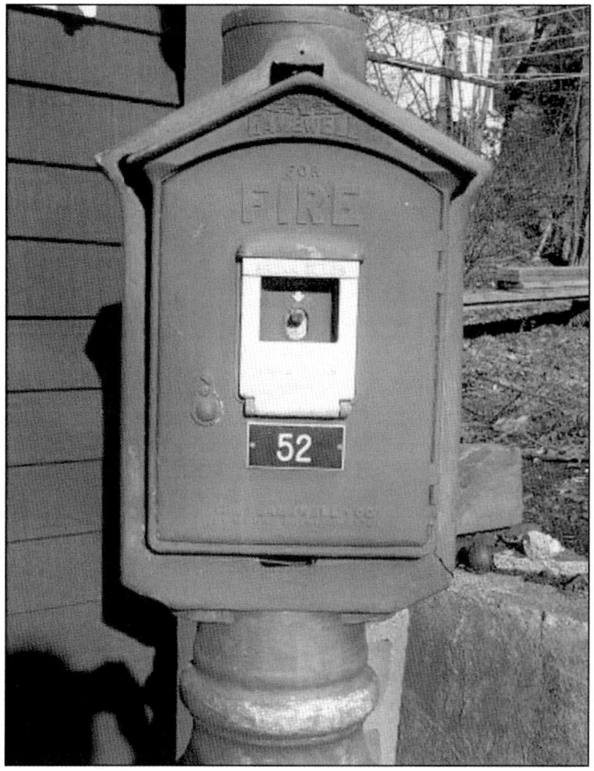

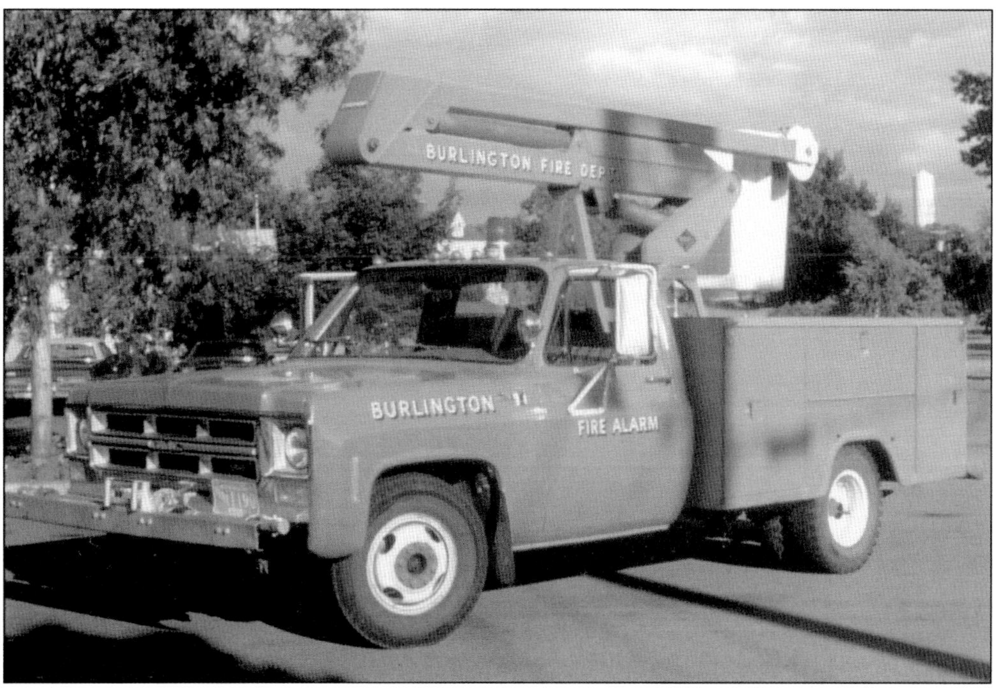

A fire alarm truck was used in the installation and upkeep of the system. When the system was installed, circuitry and wiring connected the station to alarm boxes everywhere. Today the alarm system has been upgraded and brought up to date. (Courtesy of Albert Fay.)

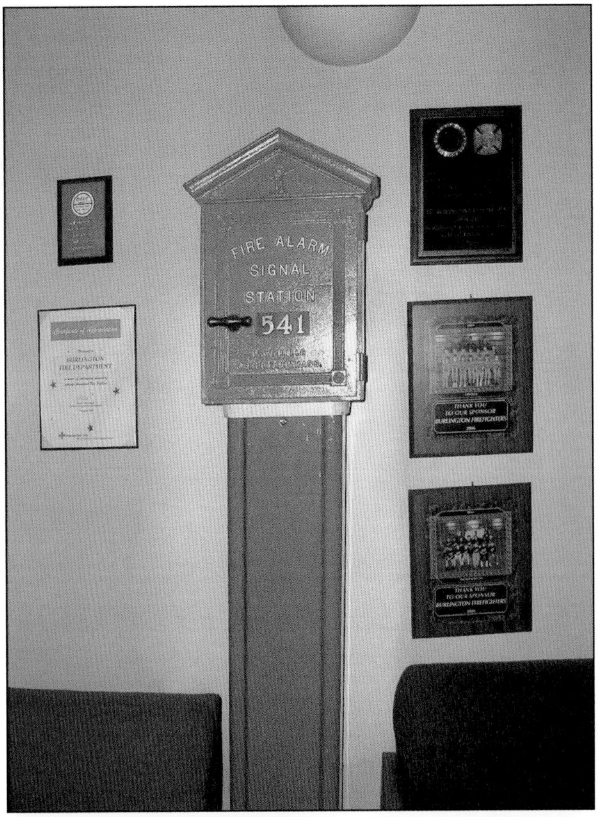

The original fire alarm box from the Church of Christ Congregational on Lexington Street was one of the first boxes put into service. It is now on display in the lobby of the fire headquarters. (Courtesy of Toni Faria.)

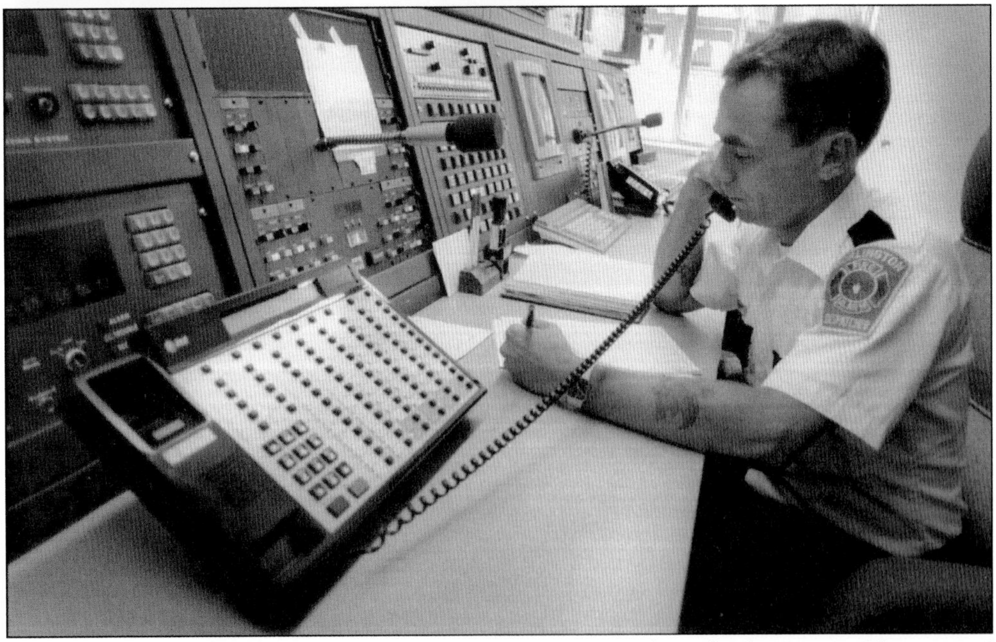

Firefighter Michael Runyan is shown here at the present console at the fire station. The original fire alarm met the needs of that time. The present-day system has evolved into a technical and modern system to keep abreast of the changing fire service. (Courtesy of Albert Fay.)

Five

ADVANCEMENT AND DEVELOPMENT OF THE DEPARTMENT

From 1960 to the present time, education and training have been essential to advancement of the fire department. In 1962, engine 6, a B model Mack, was delivered as well as a new chief's car. Equipment and workforce increased. There were now 19 full-time men and a new ambulance, as well as a new chief's car. The following year, the firefighters responded to 1,019 calls. In three years, the number of full-time men jumped to 27. By 1969, there were 36 permanent men and 10 call firefighters. The department continued to experience changes and to educate the public about fire prevention, CPR, and fire safety. Demonstrations were conducted regularly at the Burlington schools, the Burlington Mall, and the Lahey Clinic. New building codes, the use of fire alarms and sprinklers, and an educated public brought about a decrease in house fires. The constant growth of the town brought about an increase in rescue and emergency situations. In-service inspections continued to be important to maintain fire safety.

The inspections were essential to maintain fire safety for local businesses, and the fire department gained information about the companies and buildings.

Fire Chief Herbert Crawford retired in 1985. He was followed by Lawrence Rice as acting chief, then Paul Thibault, and then Lee Callahan, the present chief.

Clearly Burlington firefighting was striving to keep both residents and the daytime population employed in Burlington safe and secure.

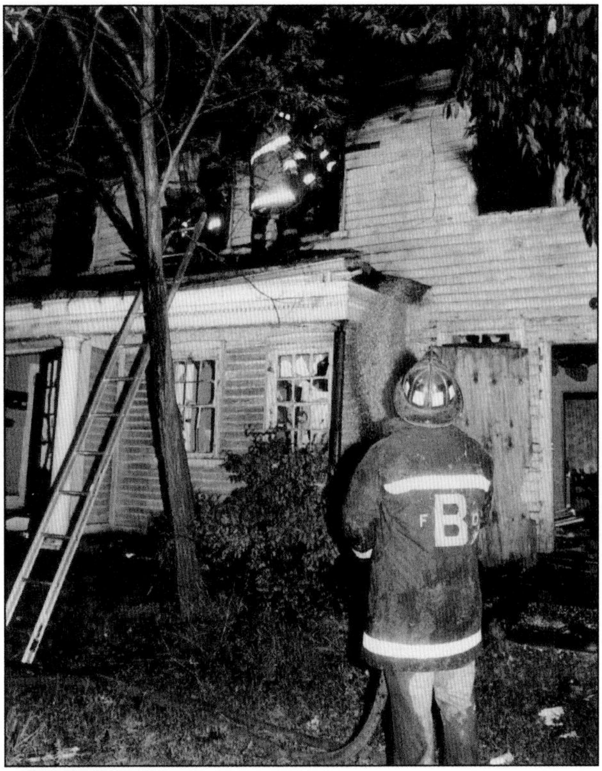

In 1983, a fire in the Frothingham mansion started in the front staircase of the house and climbed up in the center of the stairwell to the cupola. Firefighters saved the historic house, built in 1885. In 1950, it was considered the finest house in Burlington. From left to right are Lawrence Rice and Chief Herbert Crawford. (Courtesy of Herbert Crawford.)

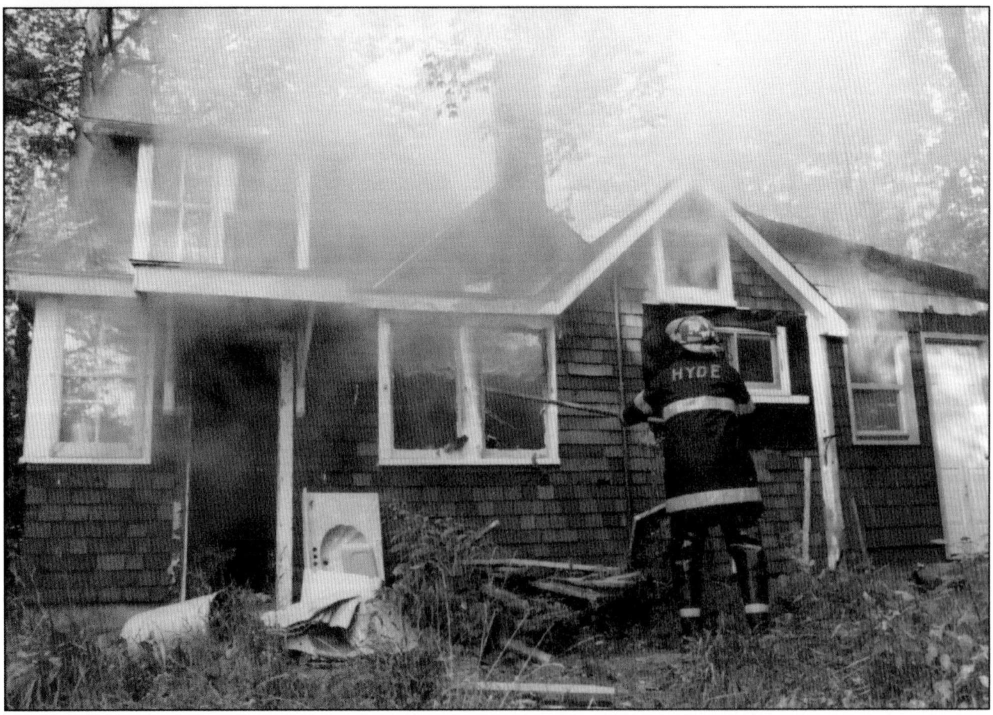

A 1987 house fire of an older wooden structure on Hillcrest Road was quickly extinguished by the firefighters. Firefighter Douglas Hyde helped to control the fire. (Courtesy of Lawrence Rice.)

Chief Herbert Crawford (right) and Gerald Mills assess the damage at the Church of Christ Congregational on Lexington and Bedford Streets. Originating in the stairwell of the addition, the fire spread, damaging the building. (Courtesy of Herbert Crawford.)

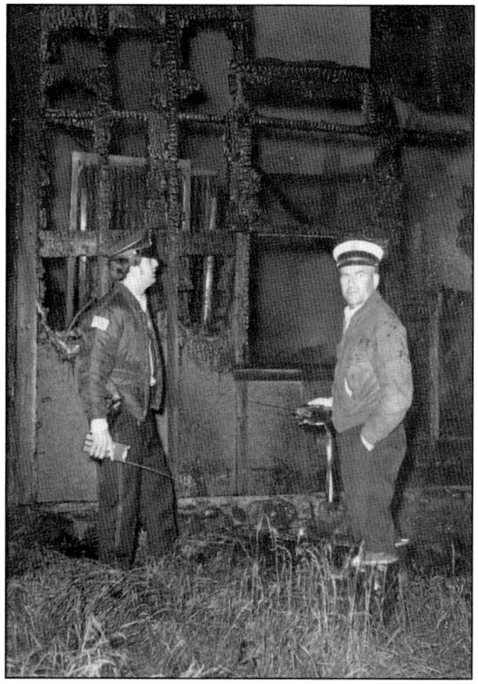

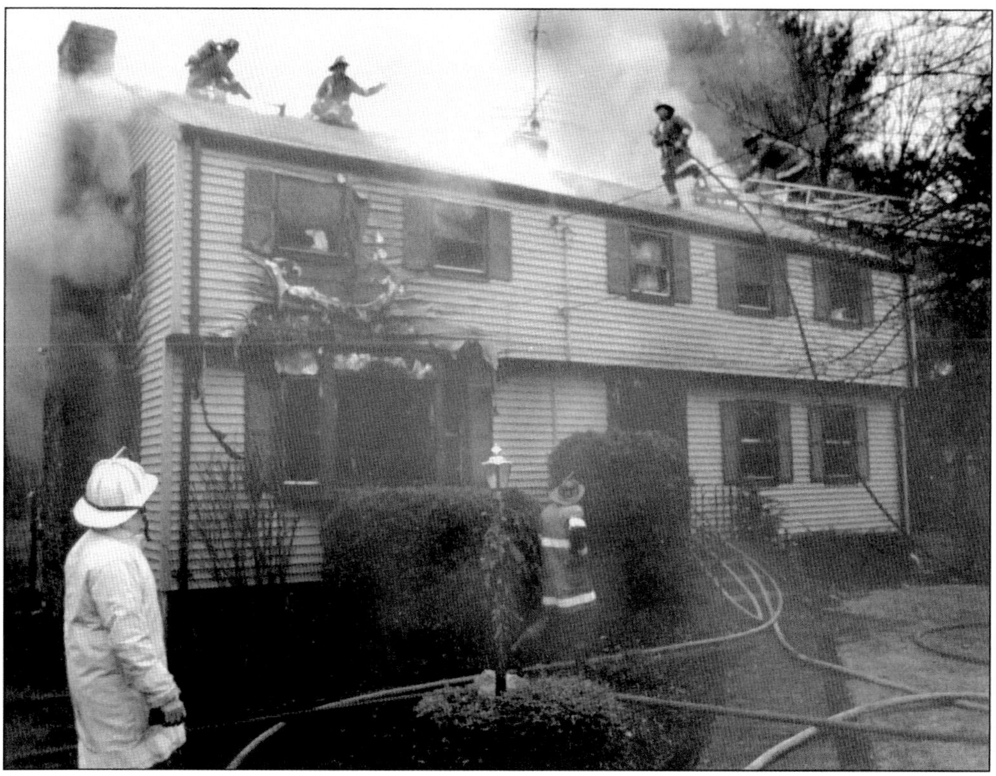

A fire at 15 Phyllis Avenue in 1989 shows the firefighters moving across the roof to ventilate the two-alarm house fire. All occupants escaped. (Courtesy of Albert Fay, photograph by Kapp Photos.)

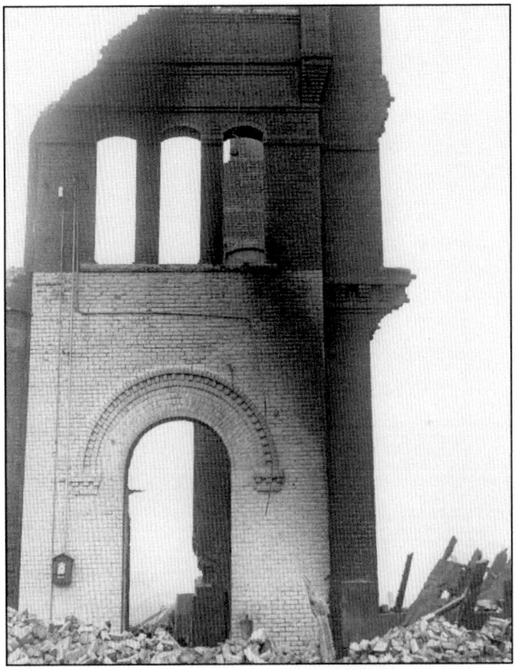

A historic fire in Lowell, the devastating Lawrence Mill complex fire in 1987, was fought by the firefighters of 17 towns. Burlington's fire department responded and helped to fight the fire that took nine hours to contain. It was an unforgettable firefighting experience for all who were working to extinguish the fire. (Courtesy of Lawrence Faria.)

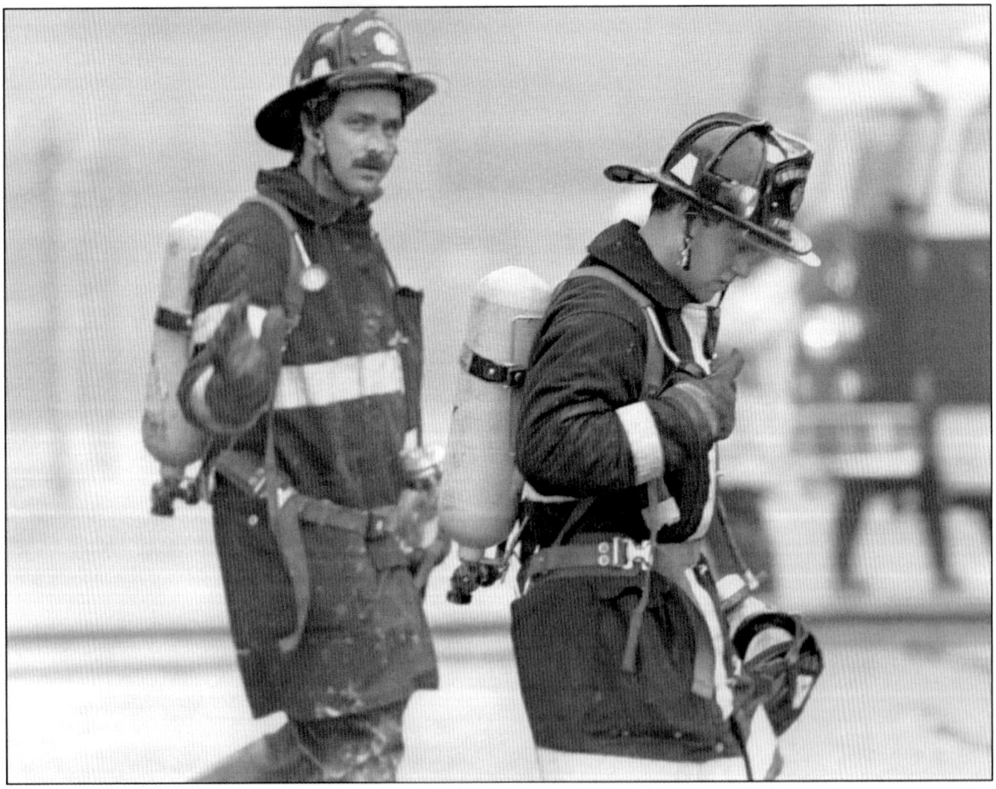

Firefighters David Sullivan (left) and Timothy Browne leave the High Voltage Corporation fire in 1991. A full complement responded to the fire. Venting through the roof was necessary to remove materials from the ceiling. (Courtesy of Albert Fay.)

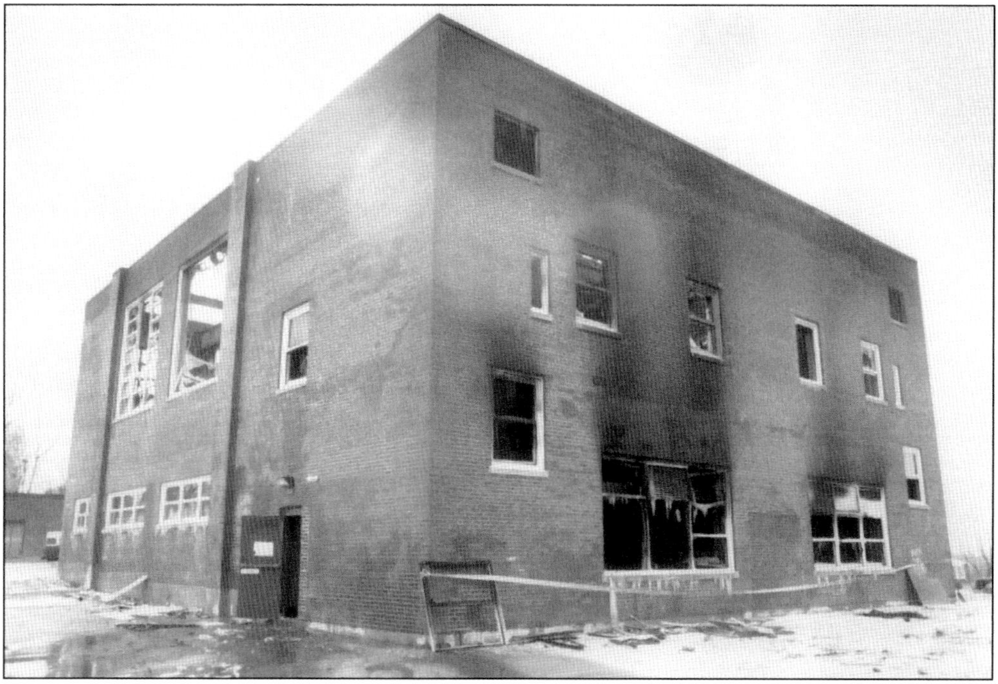

A fire occurred at the Center School, located in the center of town, during the construction of the building in February 1989. The damage is evident. (Courtesy of Albert Fay, photograph by Joe Brown.)

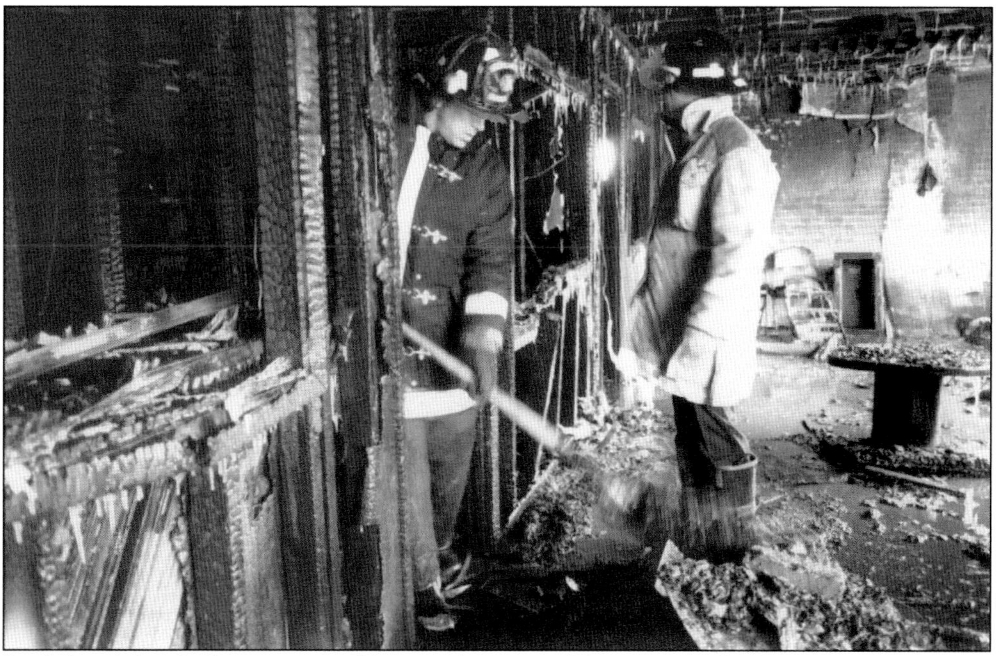

Firefighter John Corbett is pictured here at the scene of the Center School fire in 1989. This was a general-alarm fire that caused extensive damage to the building. The school was rebuilt. Later it became the recreation department and eventually the Center for Human Services. (Courtesy of Albert Fay, photograph by Marc Howland.)

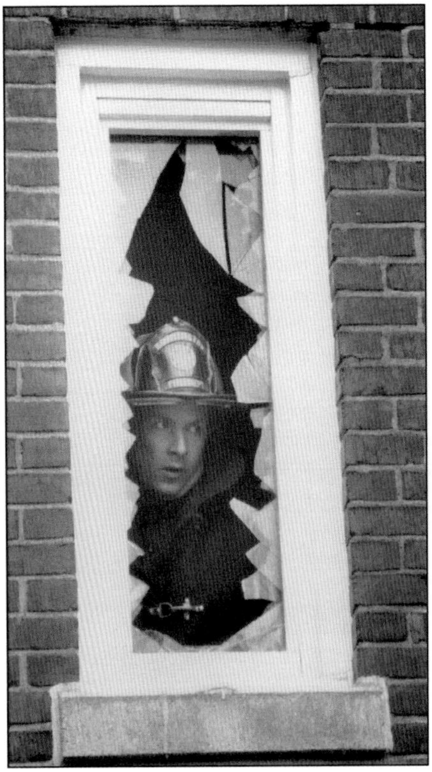

At the Center School fire in 1989, firefighter Lee Callahan is seen looking out of the gymnasium window of the school. He is currently the fire chief in Burlington. The building is still in use. (Courtesy of Lee Callahan.)

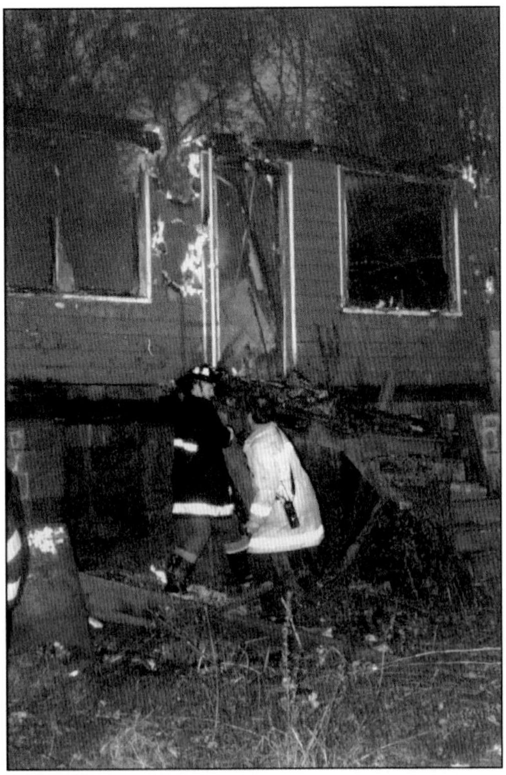

Firefighters Gary Arbing (left) and Phillip Pollicelli arrive at a house fully involved on Kenmere Avenue in 1990. (Courtesy of Burlington Fire Department.)

Lt. George Creamer is shown at the Partridge Lane fire in 1993. Lt. Lee Callahan was off duty and across the street from the scene when the fire broke out. He rushed in and rescued a young man from the fire. The man suffered burns over 90 percent of his body. The incident and shift commander was Capt. Edward Costa. (Courtesy of Burlington Fire Department.)

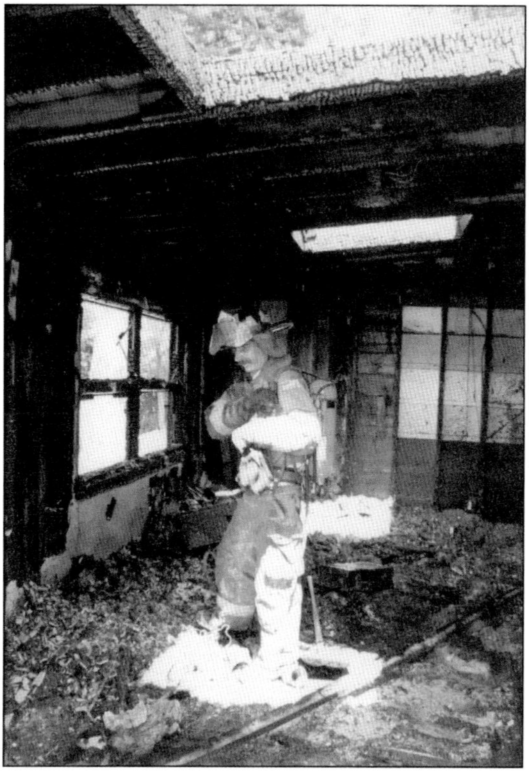

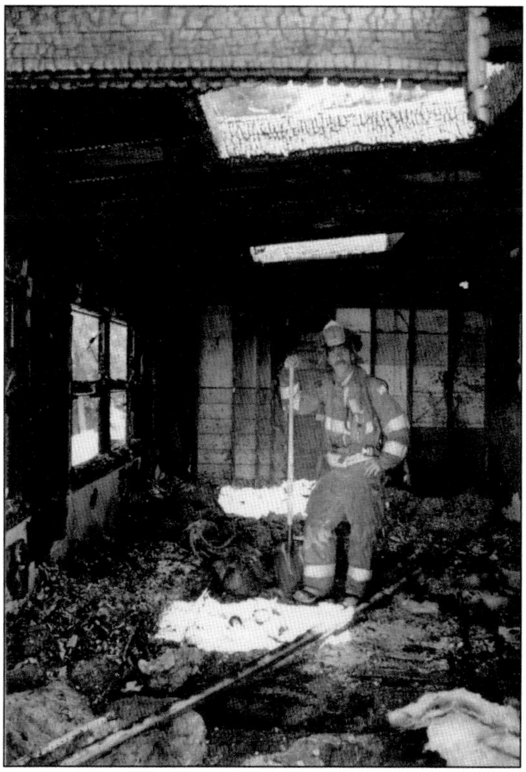

This is a dramatic image of George Creamer cleaning up after the fire was brought under control at the devastating Partridge Lane fire. (Courtesy of Burlington Fire Department.)

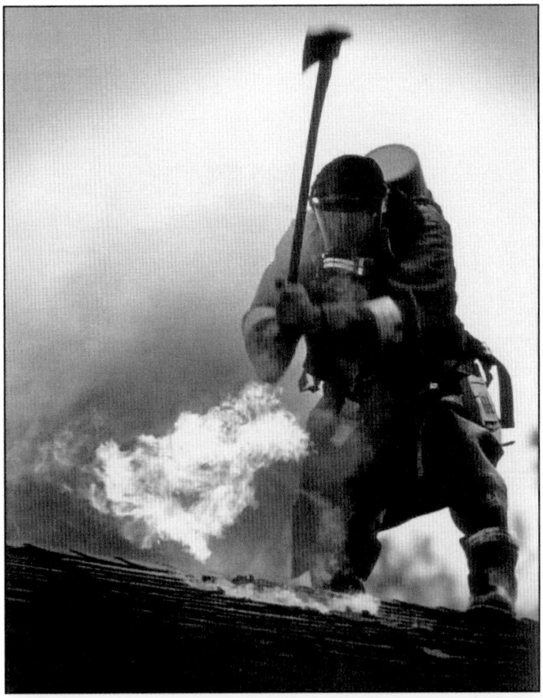

Firefighter Douglas Hyde wields an ax to vent the fire on the roof of a house fire. The date and site of this photograph are unknown. (Courtesy of Albert Fay.)

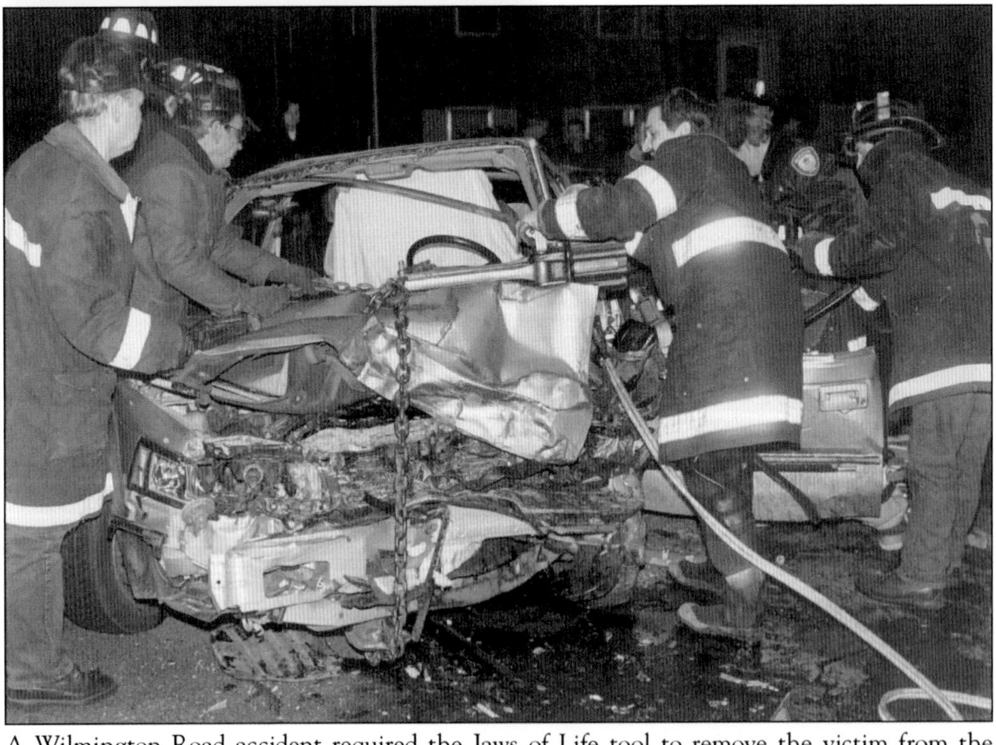

A Wilmington Road accident required the Jaws of Life tool to remove the victim from the vehicle. From left to right, William Lundgren, Vincent Pollicelli, and Edward Costa operate the tool. The tool had been donated by a local fund-raising group. (Courtesy of Albert Fay, photograph by Joe Brown.)

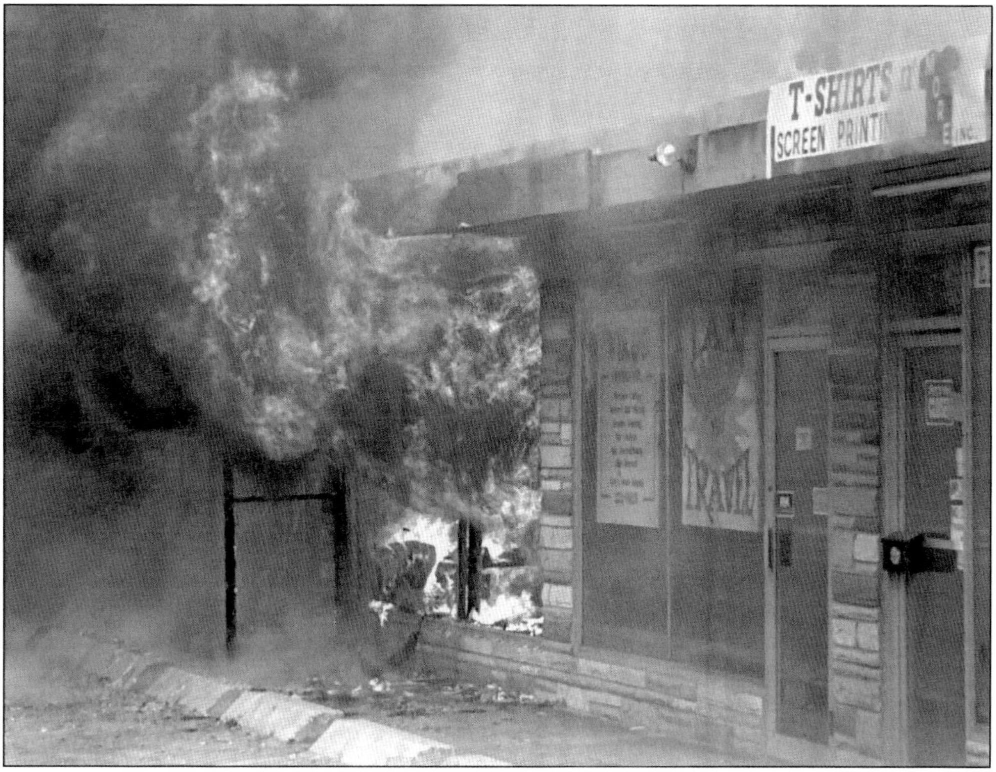

The front windows are being blown out of the Winn Street retail block fire before firefighters arrive at the fire scene in 1987. (Courtesy of Albert Fay, photograph by Joe Brown.)

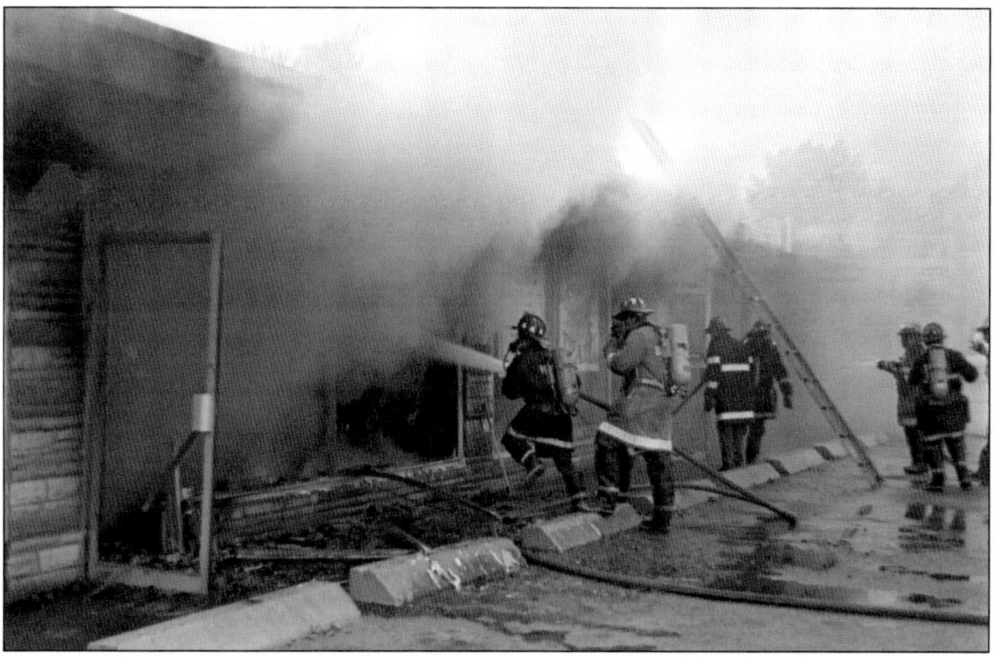

Firefighters bring the fire under control at the Winn Street retail block fire that threatened the remaining stores on the block. (Courtesy of Albert Fay, photograph by Joe Brown.)

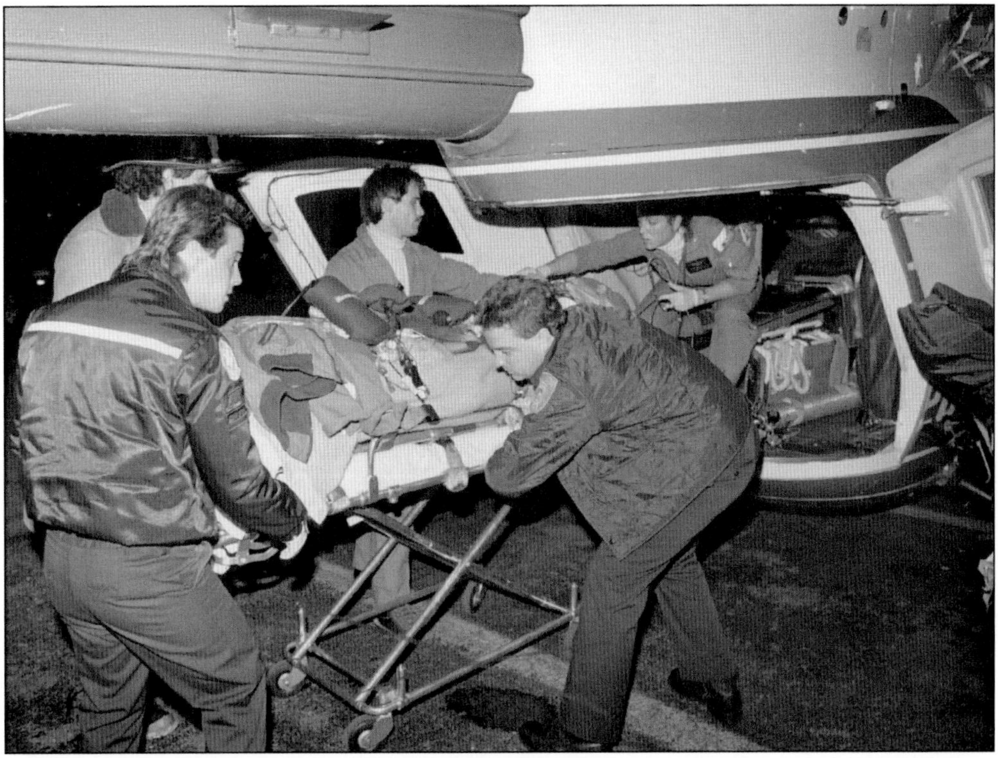

A rescue at the Mall Road and Middlesex Turnpike involved moving the victim onto the New England Medical Flight. From left to right, firefighters Albert Brabant, Michael Bibbey, Paul Sullivan, and Gerard Letendre are assisting. The date is unknown. (Courtesy of Albert Fay.)

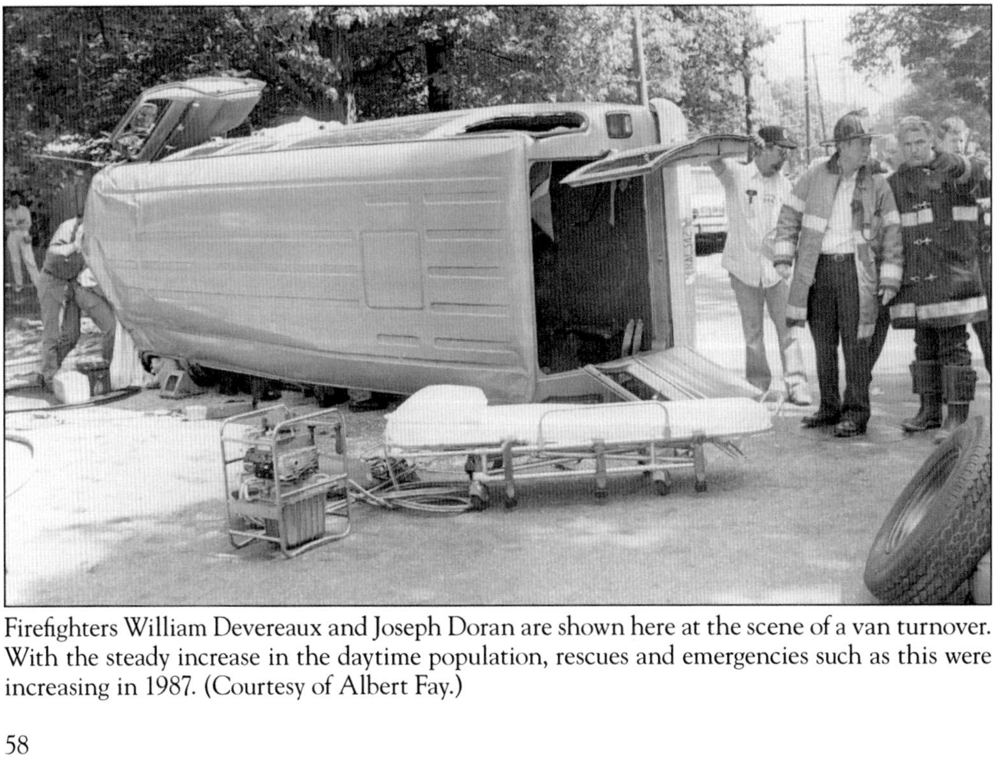

Firefighters William Devereaux and Joseph Doran are shown here at the scene of a van turnover. With the steady increase in the daytime population, rescues and emergencies such as this were increasing in 1987. (Courtesy of Albert Fay.)

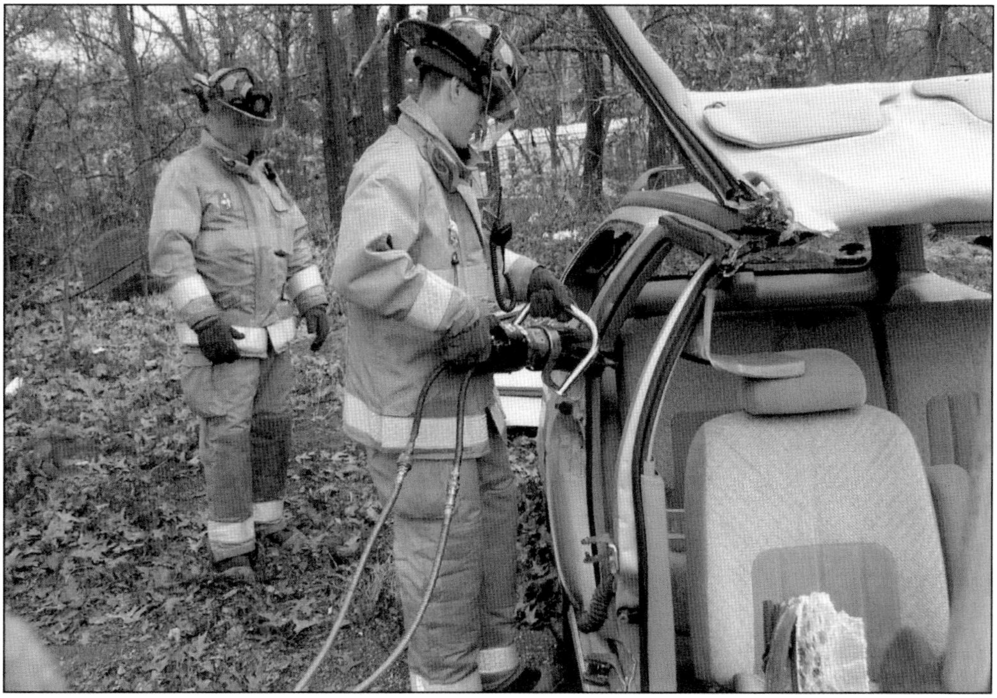

Due to the increase of motor vehicle accidents, today's firefighters require special rescue training to free trapped motorists. Firefighter Robert Paul is shown using the rescue tool. (Courtesy of Burlington Fire Department, photograph by Michael Patterson.)

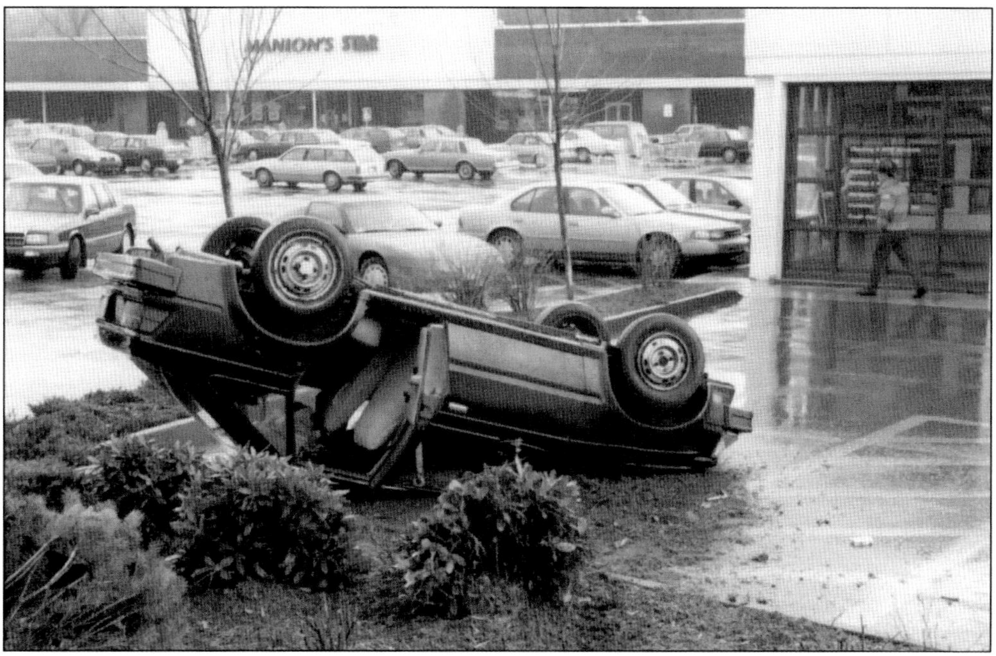

In 2005, firefighters were called to an automobile turnover at Bedford Street and Middlesex Turnpike, a heavily traveled road. More emergencies and rescues replaced house fires as fire prevention education became more widespread. (Courtesy of Paul McLaughlin.)

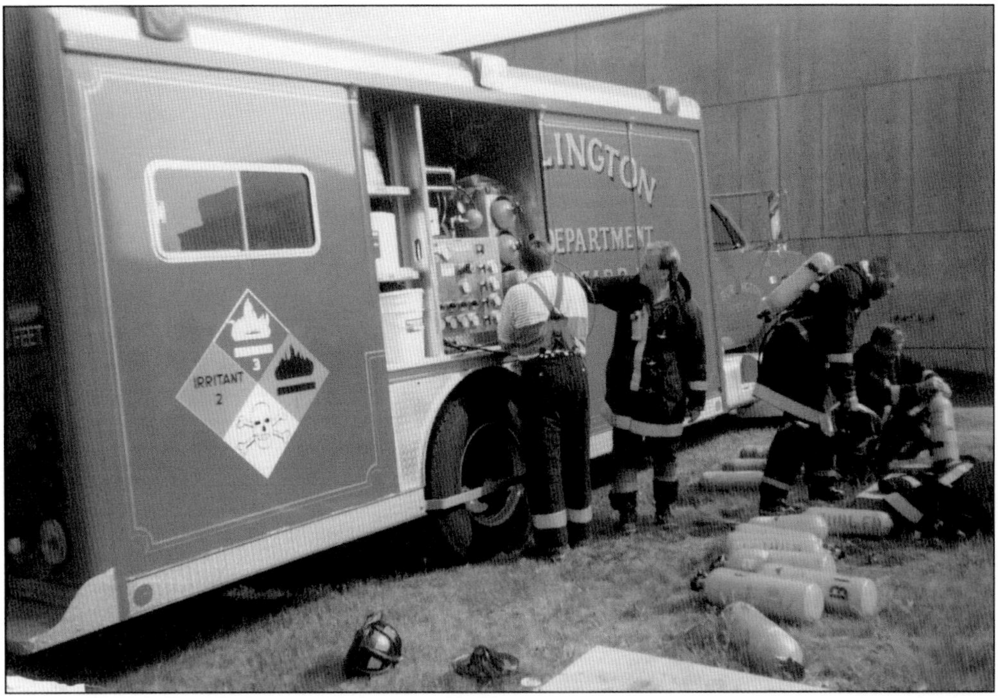

Pictured here is the Burlington High School hazardous materials incident. John Corbett, Michael Patterson, Frederick Williams, and Paul McLaughlin, shown here from left to right, responded to the call. (Courtesy of Burlington Fire Department.)

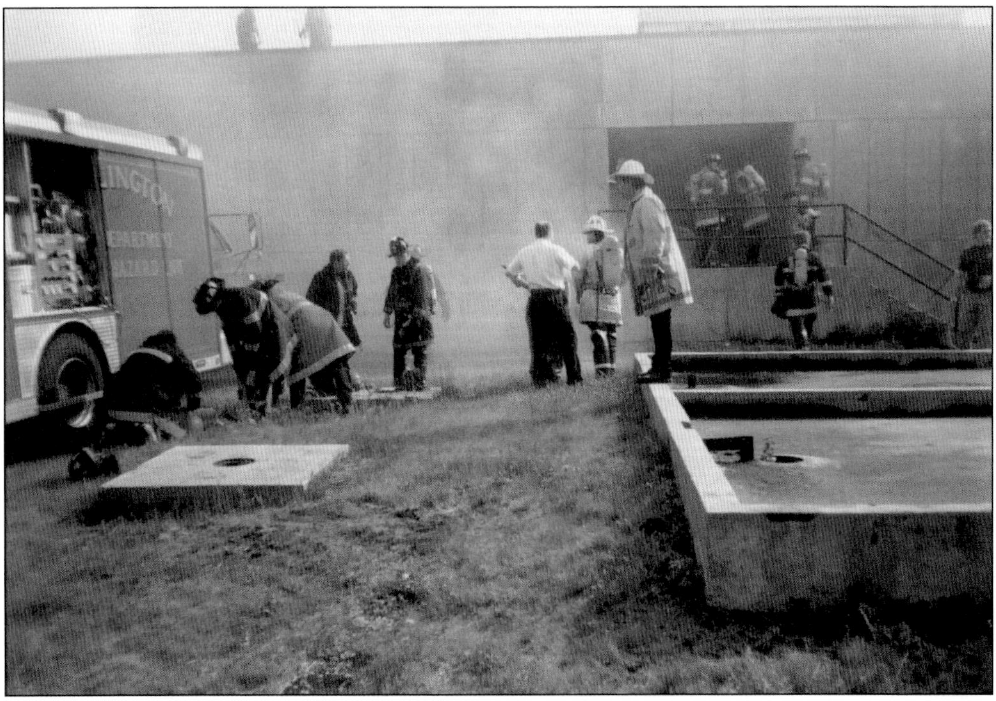

The hazardous materials vehicle responded to the emergency at the high school. Deputy Chief Phillip Pollicelli is seen in the foreground. (Courtesy of Burlington Fire Department.)

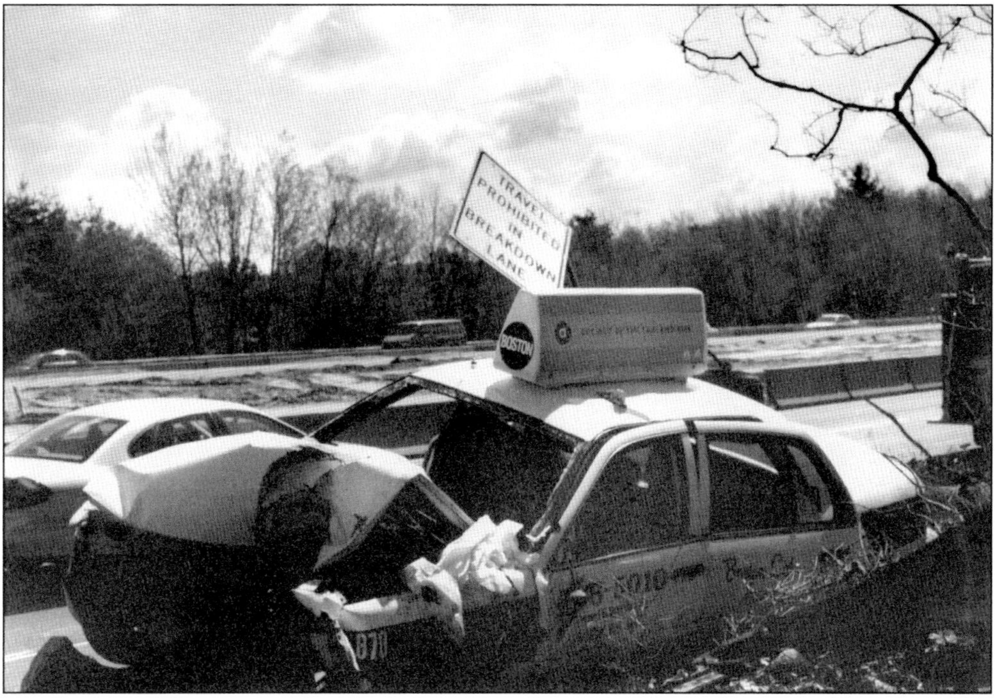

An accident in the vicinity of a Route 3 construction site involved two vehicles. Both vehicles went off the road. An increase in the daytime population resulted in an increase in accidents and emergencies of this type. (Courtesy of Paul McLaughlin.)

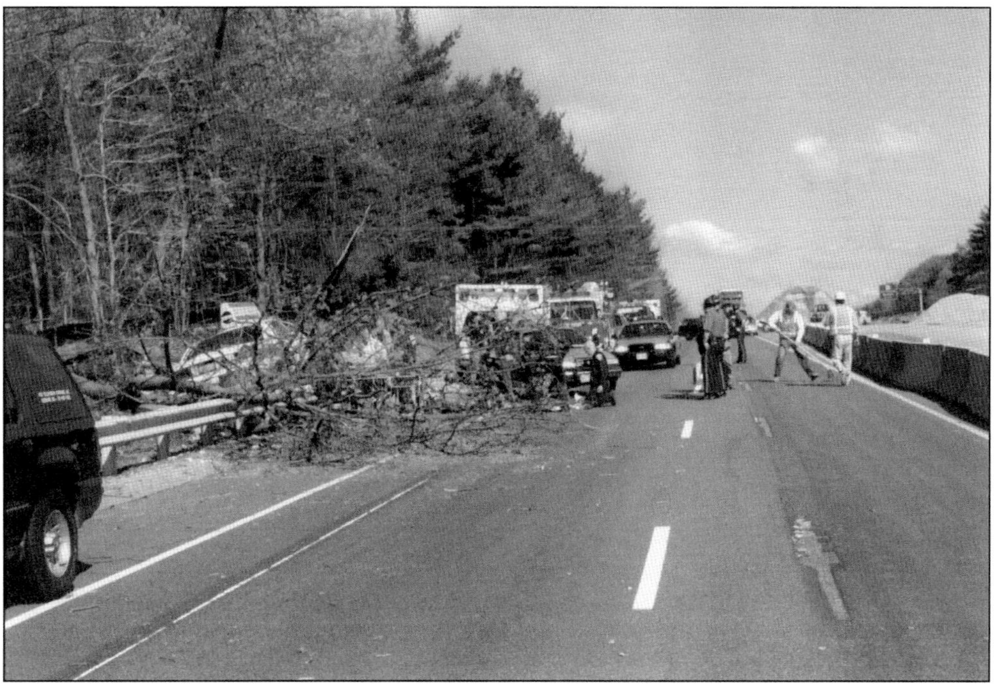

The accident on Route 3 illustrates the damage to the roadway and vehicles as well as the subsequent delays in traffic. (Courtesy of Paul McLaughlin.)

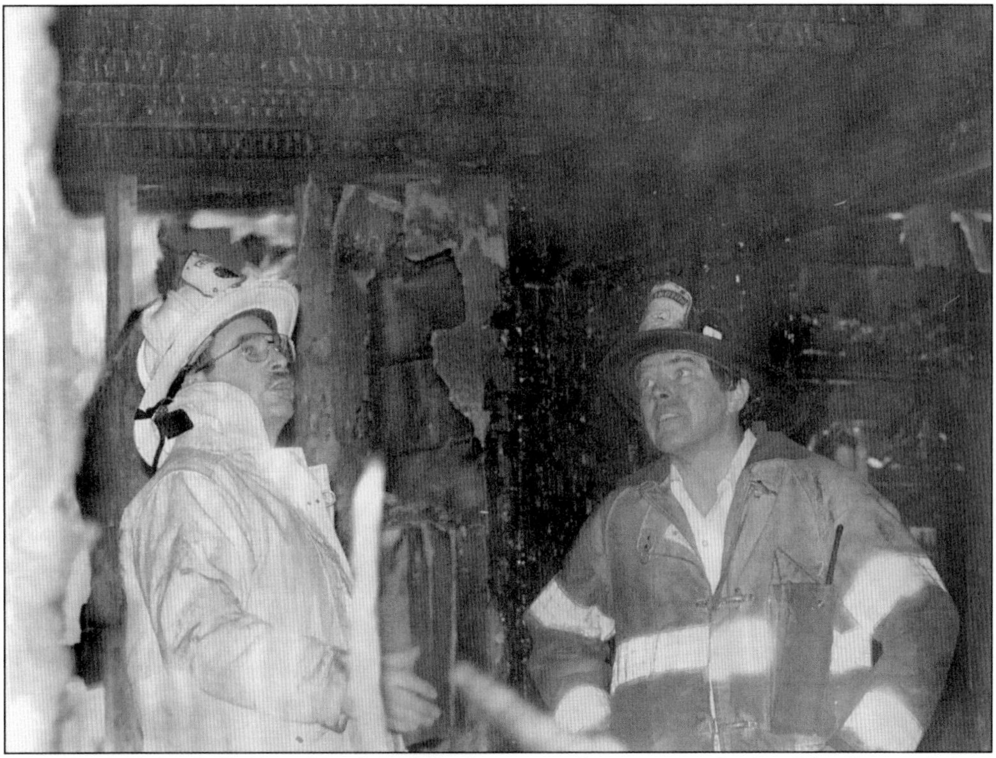

Chief Paul Thibault (left) and Paul Gerbrands inspect the remains of a fire. The date and place are unknown. (Courtesy of Paul Thibault.)

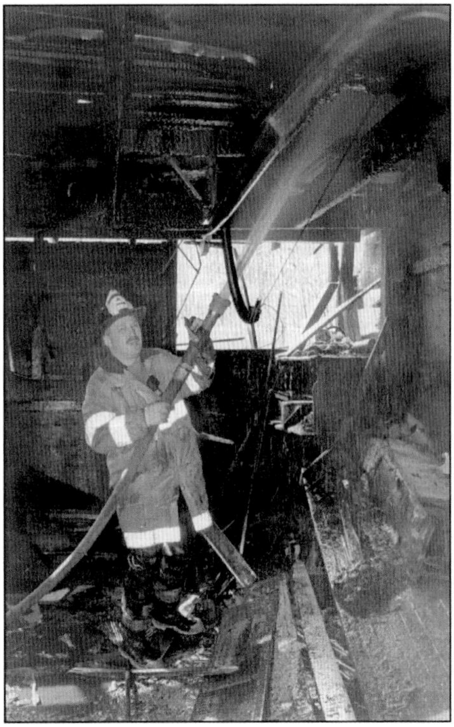

Firefighter Donald Robinson fights a garage fire. The date is unknown. (Courtesy of Albert Fay.)

Herbert Crawford retired in 1985 after 30 years of service. Chief Crawford kept the department abreast of the needs of the town, which grew from a farming community of 3,000 at the end of World War II to a thriving community of 24,000. The task was accomplished with excellent results. He was voted Outstanding Citizen of the Year in 1986. (Courtesy of Karen Crawford Pflanz.)

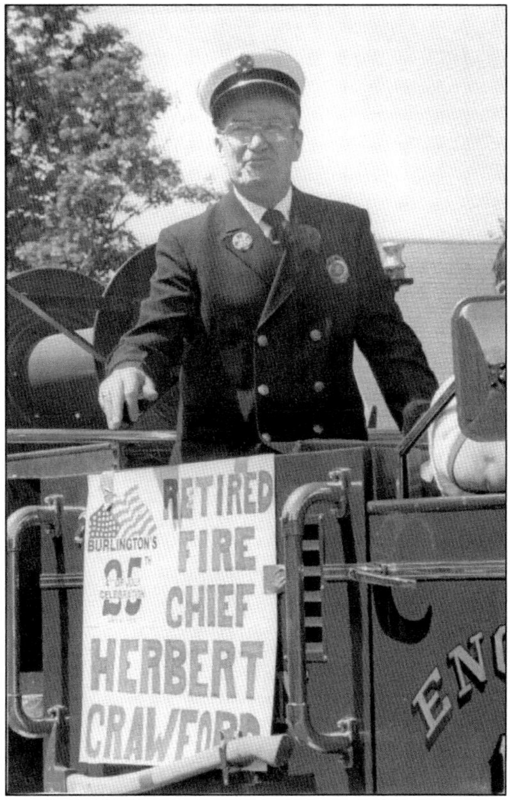

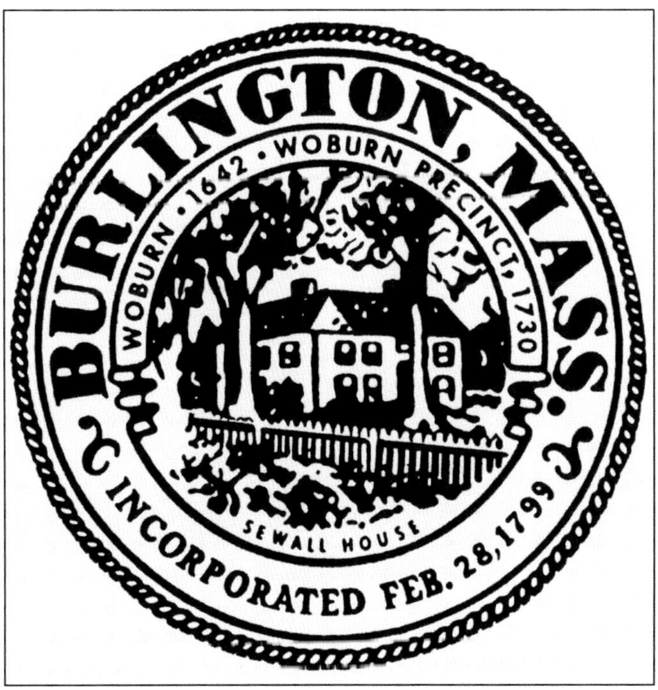

The town seal of Burlington features an image of the Sewall House, one of the earliest examples of a historic property lost to a fire. (Courtesy of Burlington archives.)

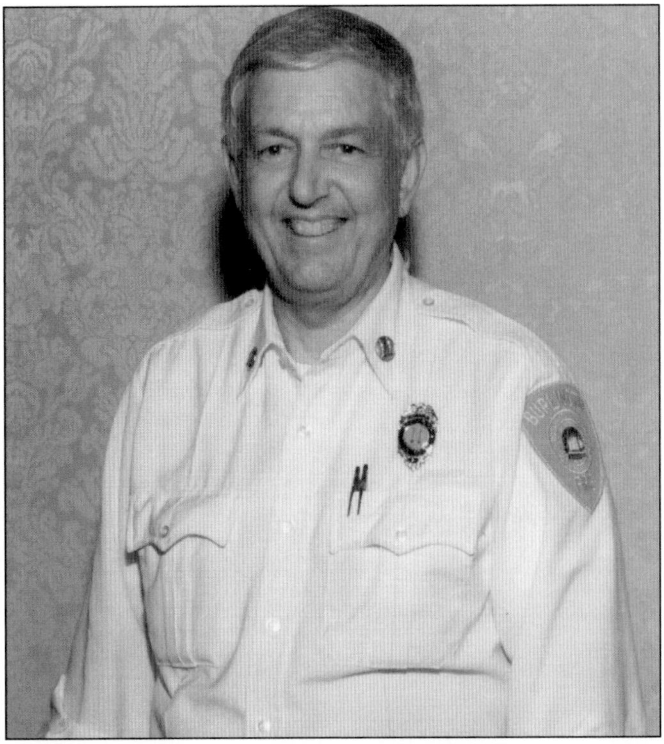

Capt. William Devereaux began as a call firefighter in 1969, and in 1970, he was appointed a full-time firefighter. For many years, he coordinated all fire suppression training and EMT and CPR training. Captain Devereaux conducted the training to keep pace with the changes and demands of the community. Before his retirement in 2001, he worked with the Federal Bureau of Investigation reviewing terrorist tactics and high explosive situations. (Courtesy of Albert Fay.)

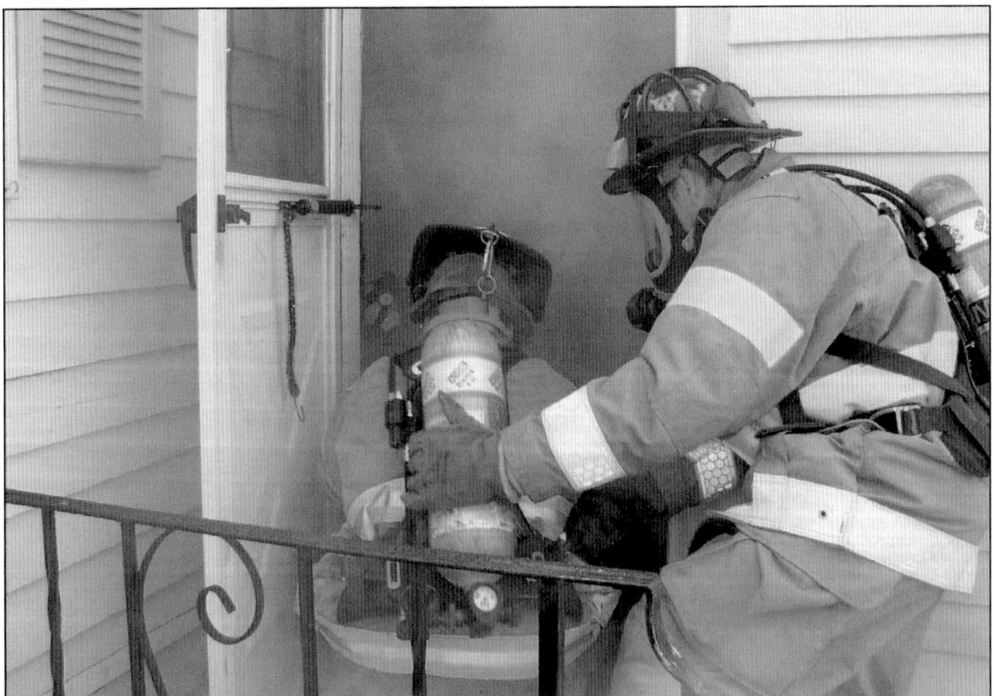

Public awareness is an important part of the fire service. The department holds open house at the station, and it hosts field trips from the elementary schools. Shown in this photograph is Capt. Michael Patterson in 1995. He later became the training officer. (Courtesy of Paul Thibault.)

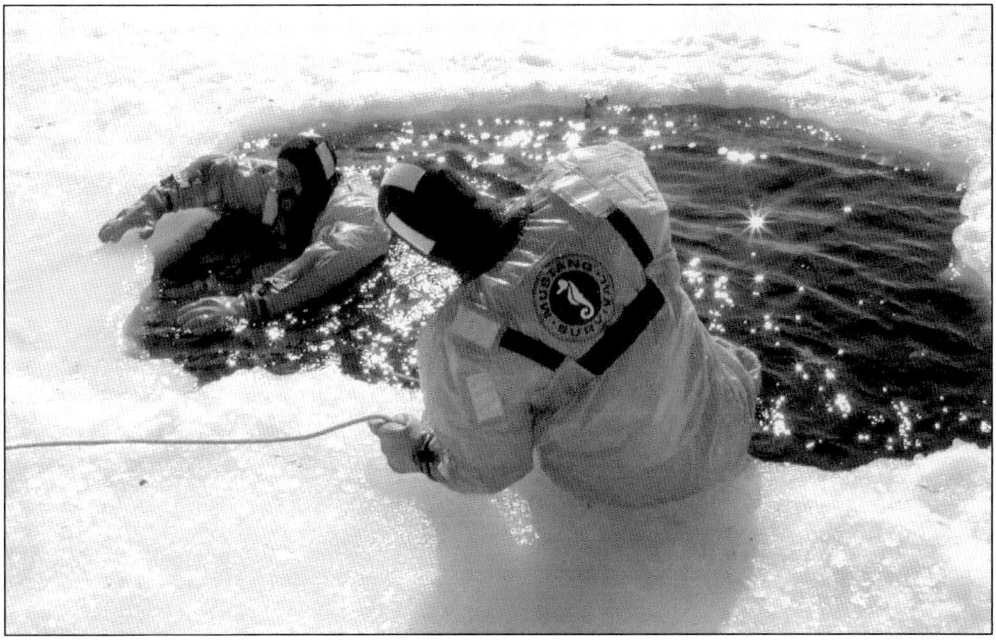

Water rescue training became part of expanded training, using new equipment to help in rescue situations at Burlington's reservoir. The education of each firefighter is essential to the success of the response to accidents and emergencies because today's firefighters are responding more often to rescues and medical emergencies. (Courtesy of Michael Patterson.)

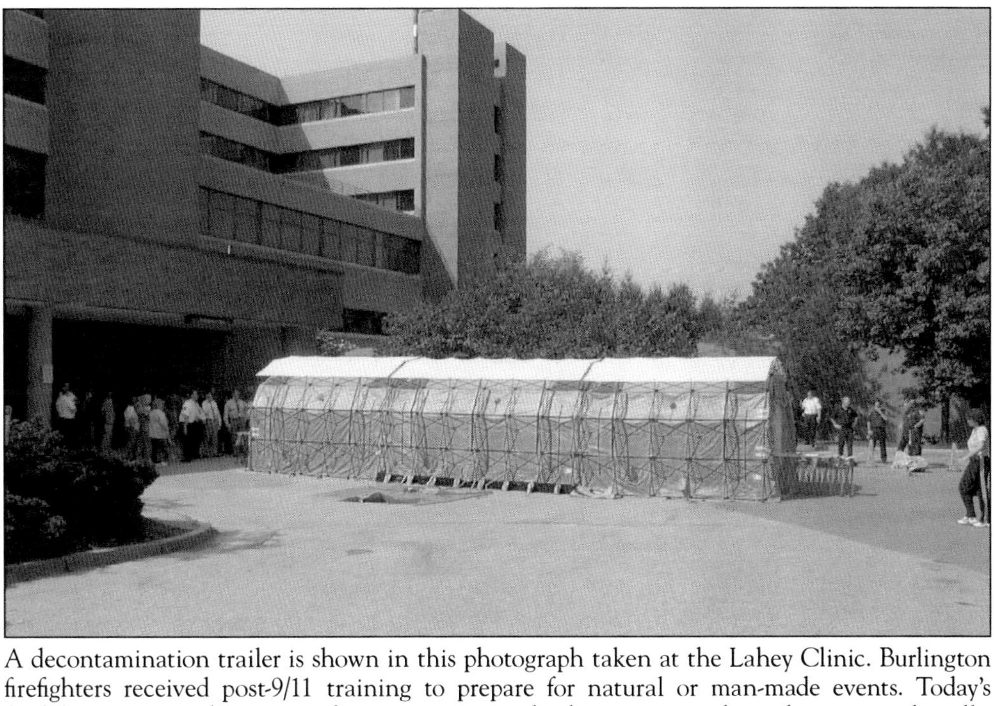

A decontamination trailer is shown in this photograph taken at the Lahey Clinic. Burlington firefighters received post-9/11 training to prepare for natural or man-made events. Today's firefighter is responding more often to rescues, medical emergencies, hazardous materials spills, and motor vehicle accidents. The constant growth of the town brought about an increase in emergency situations. (Courtesy of Michael Patterson.)

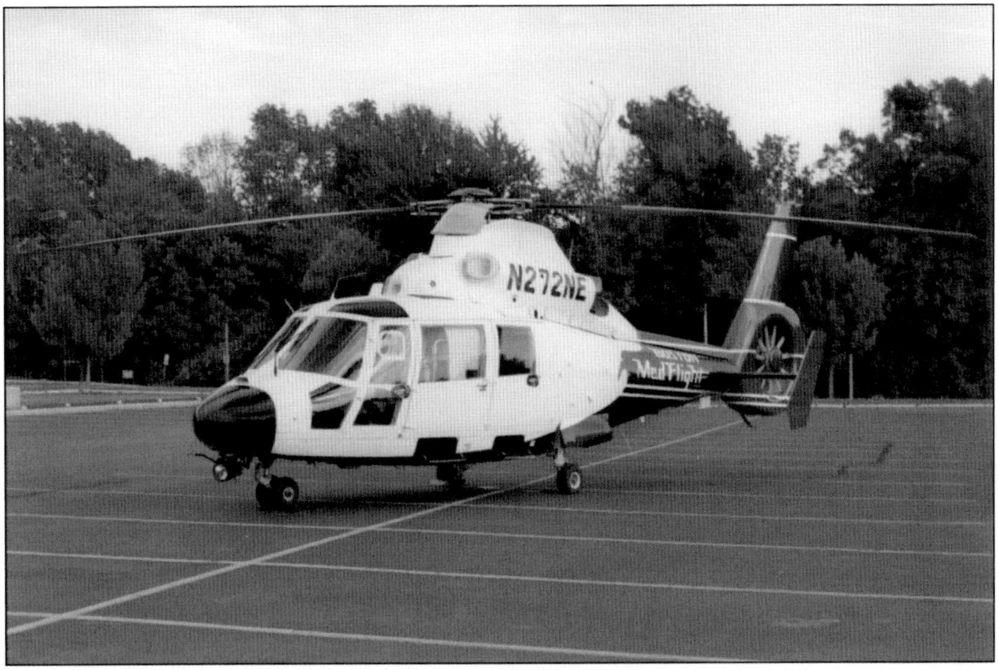

The Med Flight helicopter is based at Hanscom Air Force Base in Bedford. The Burlington Fire Department often uses the helicopter to transport patients to Boston. This service has been an asset to the department. (Courtesy of Michael Patterson.)

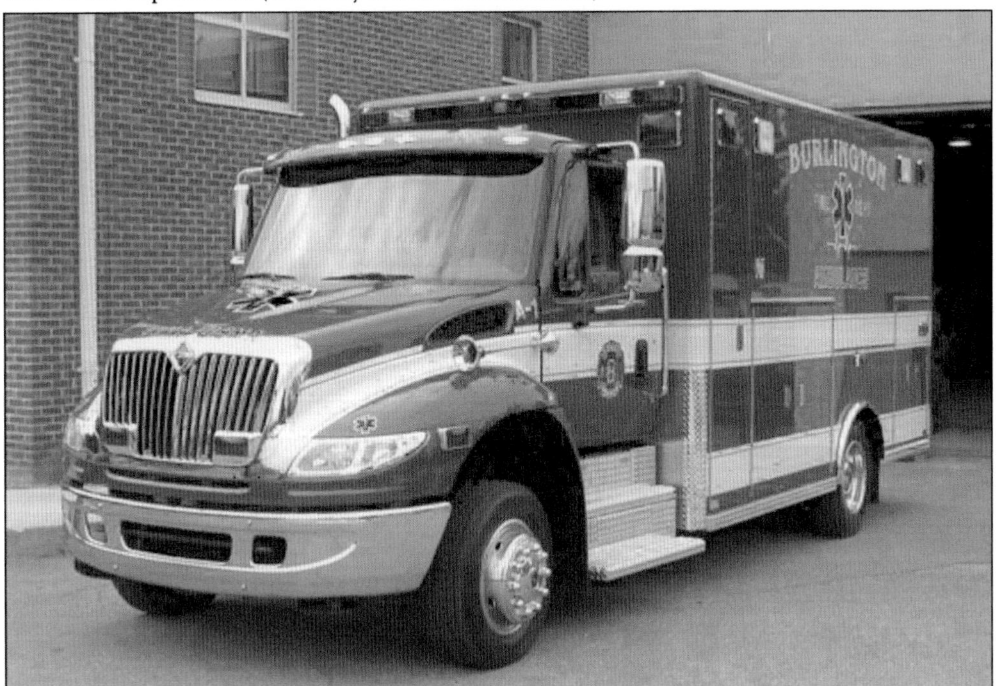

In keeping with the latest equipment and technology, the department has a 2005 International/Horton ambulance, which is housed at headquarters. It also carries air bags for vehicle extrication. (Courtesy of Brendan Miccichi.)

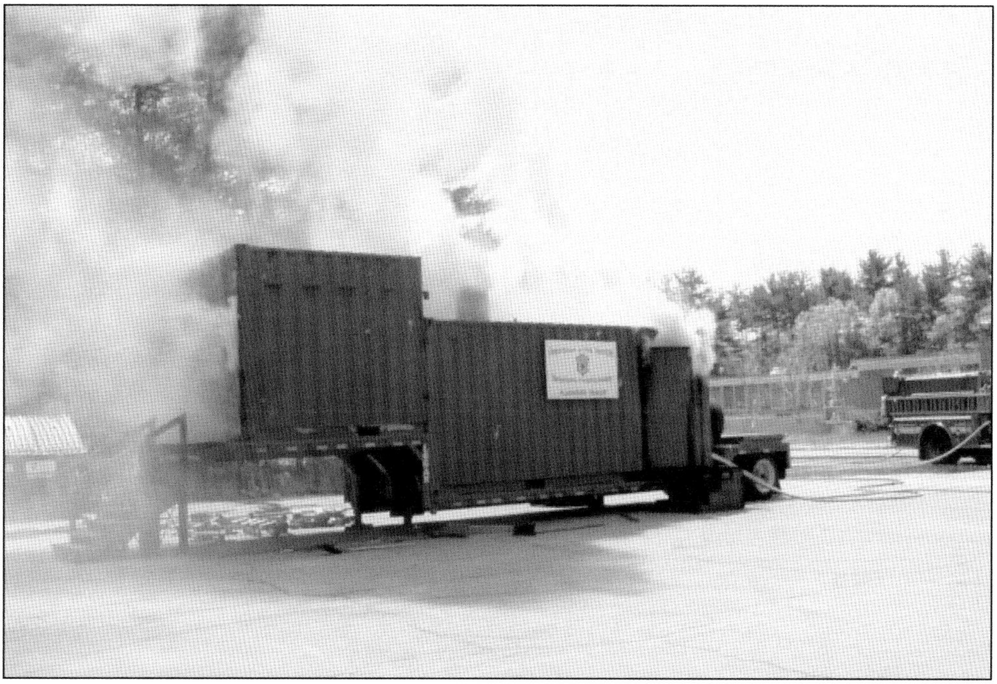

Training is ongoing in the department to keep abreast of new methods of firefighting. A flashover trailer is used to simulate a house fire that has no ventilation. Firefighters enter and experience the situation of a flashover, which often occurs in a house fire. (Courtesy of Michael Patterson.)

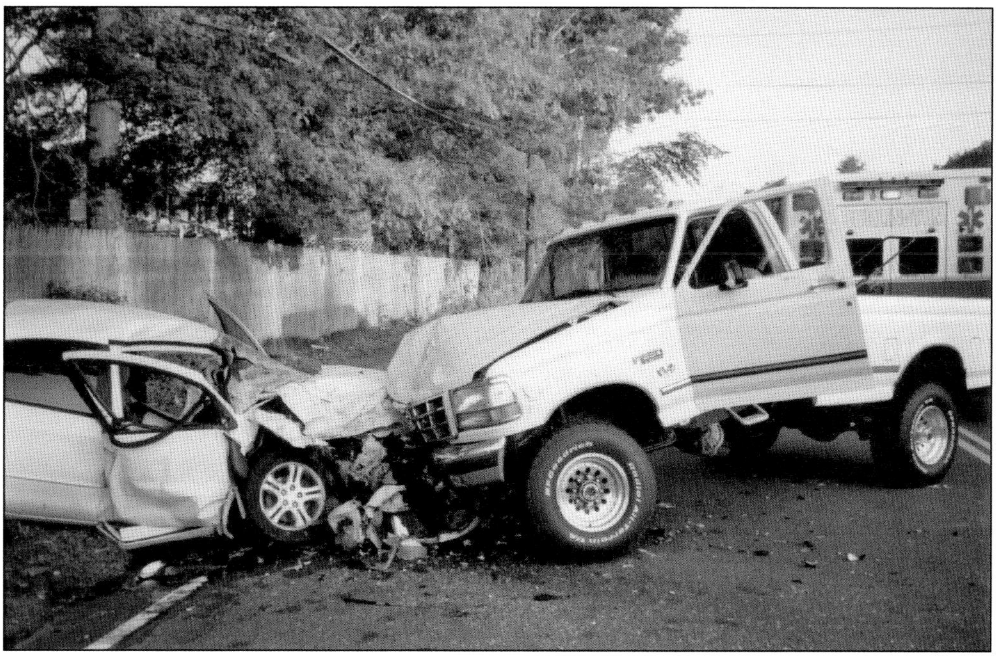

This recent motor vehicle accident is an example of the emergency and rescue calls that are now on the increase due to the daytime increase in traffic. At the same time, the number of house fires is decreasing due to smoke detectors and fire-prevention programs. (Courtesy of Burlington Fire Department.)

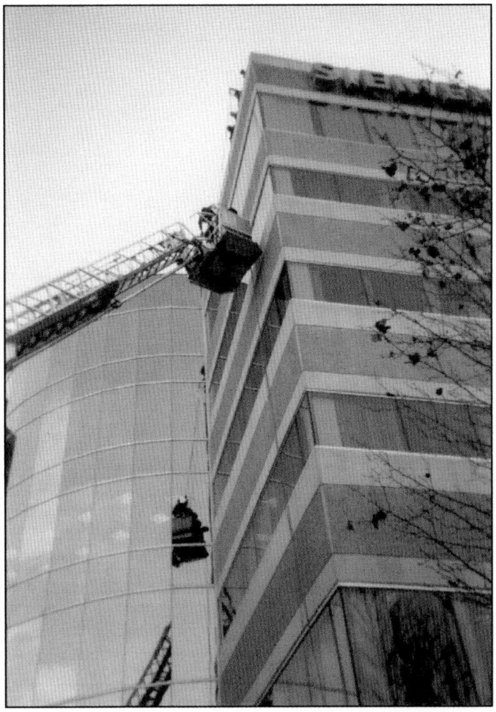

A window washer was rescued in 2004 after his equipment failed and he was left suspended on a high-rise building. The tower and experienced firefighters freed the man from this dangerous situation. (Courtesy of Burlington Fire Department.)

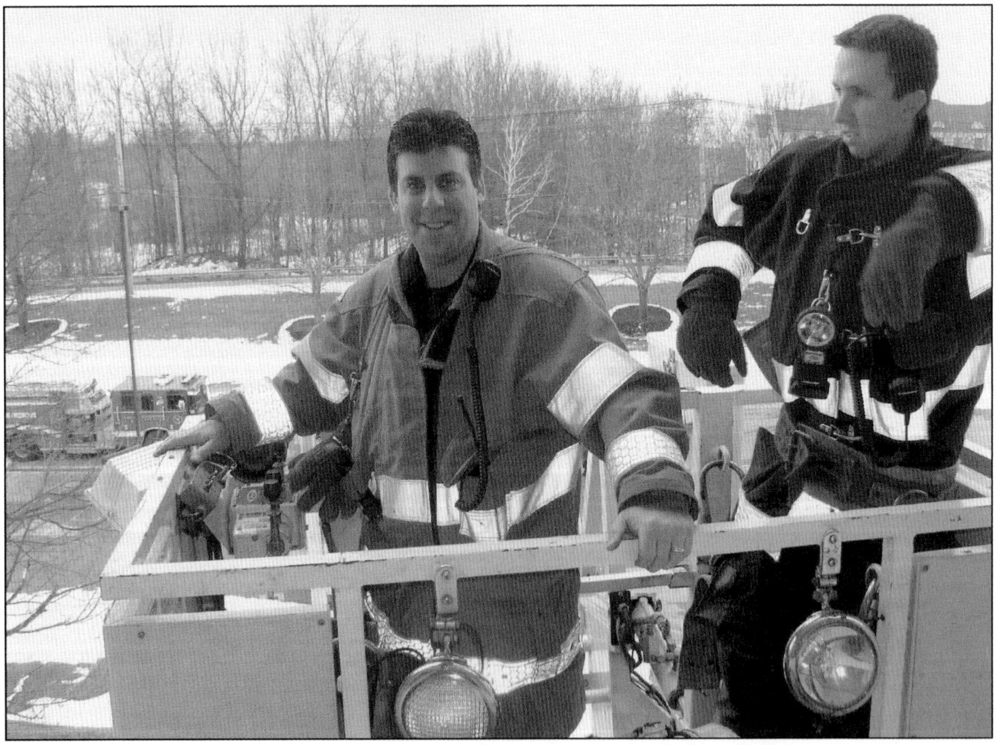

Todd Ficociello (left) and Shaun Kenney are shown during the ladder drill training. This is one example of many training exercises that take place in today's modern department. (Courtesy of Burlington Fire Department.)

Six

Firefighters Past and Present

Through the years, the town of Burlington has been fortunate to have trained men of high caliber to serve as fire chiefs. They have led effectively and consistently produced an outstanding and professional department. Each man appointed managed to add his particular expertise to the position. The fire chiefs are listed here beginning with Walter Skelton, who led the call department in 1935. Herbert Crawford was the first permanent fire chief, appointed in 1955. The men listed in this chapter have served the town of Burlington well. Thanks to the leadership provided by the officers and the caliber of the countless firefighters who are drawn to the profession, the town has a professional department that has created a better quality of life in Burlington.

The profession has attracted a second generation of firefighters. The Burlington Fire Department is honored to have eight new firefighters who were preceded by their fathers as firefighters. It is a credit to the department to have these men follow their fathers' example and become Burlington firefighters. There are also four sets of brothers and one brother and sister serving together on the department. In addition, one firefighter has a daughter who is a dispatcher on the department.

In this chapter are the fire chiefs past and present, the fathers, sons, brothers, and the two women who are related to firefighters. A collection of photographs from the past shows the parades, ceremonies, activities, and camaraderie between the men.

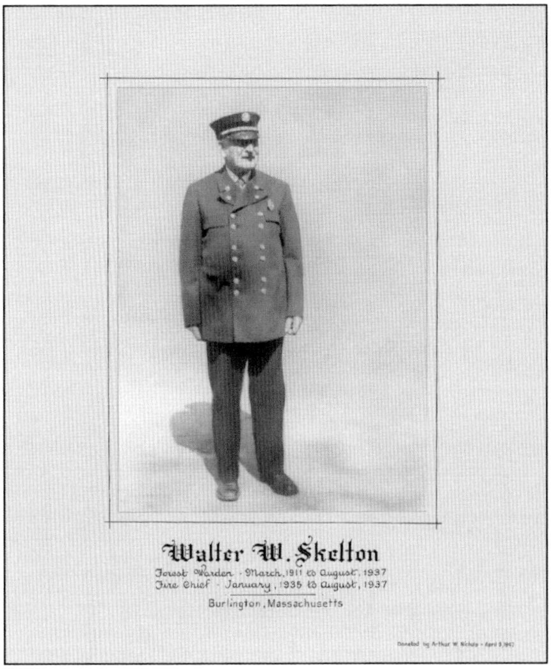

The Skelton family has been well represented in the Burlington fire service since Walter became the first fire chief of the call department in 1936. He was instrumental in the formation of the volunteer department. In 2005, the Burlington Fire Department dedicated the lobby of the fire headquarters to the many men of the Skelton family in Burlington who had been active in firefighting through the years. (Courtesy of Burlington archives.)

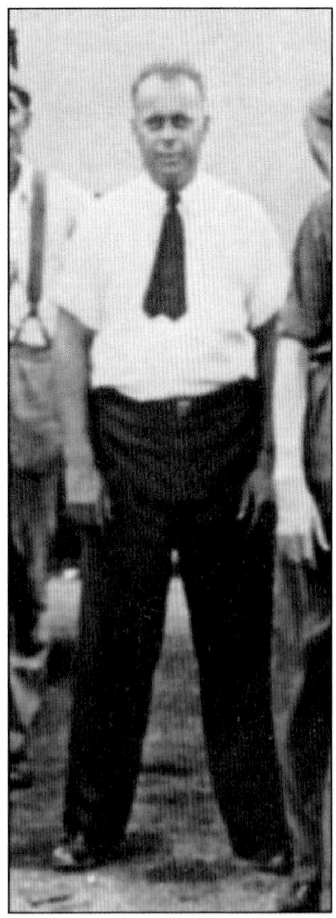

George Martikke was chief from 1937 to 1938. Chief Martikke was a call firefighter in Burlington for many years. He served briefly as the fire chief. (Courtesy of Burlington archives.)

Chief David Skelton, son of Walter Skelton, was chief from 1938 to 1951. David Skelton followed his father's example and became a leader in the fire service for 13 years. In addition, he was civic minded and served as selectman, assessor, cemetery commissioner, and a member of the school committee. He is pictured here on the left with Lester McInnis. (Courtesy of Burlington archives.)

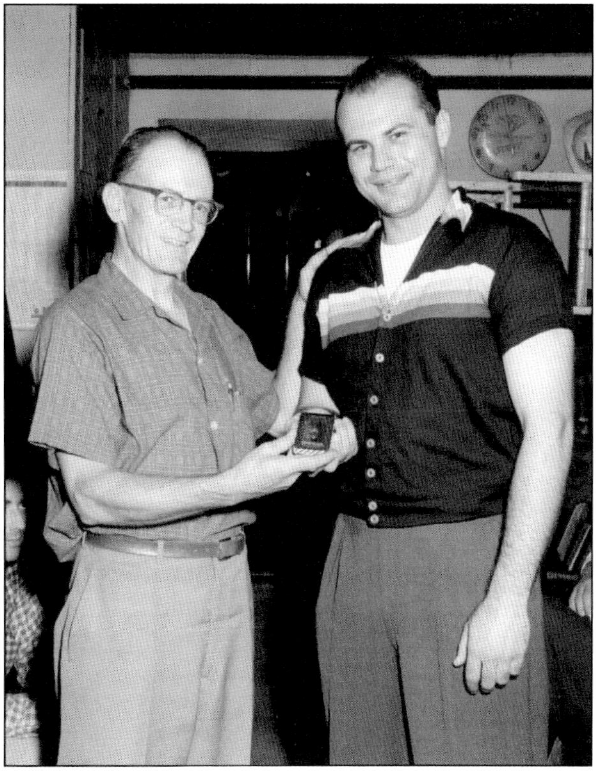

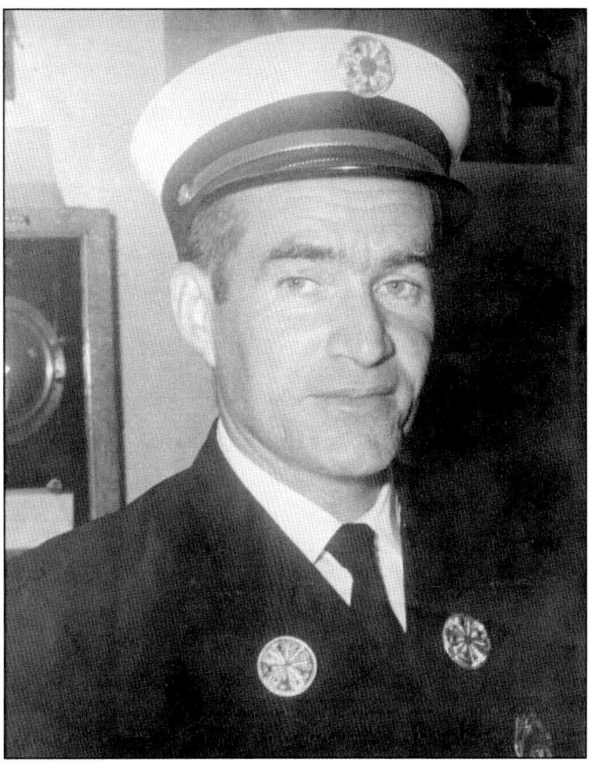

Chief Herbert Crawford was the first full-time fire chief; he was appointed in 1955 and served until his retirement in 1984. Historian John Fogelberg, author of *Burlington, Part of a Greater Chronicle*, stated in his book, "Chief Crawford brought the fire department to a state of technical excellence probably never dreamed of by Walter Skelton." His career spanned the years of growth and development in Burlington. He was chosen by the town in 1986 as the Outstanding Citizen of the Year. (Courtesy of Burlington archives.)

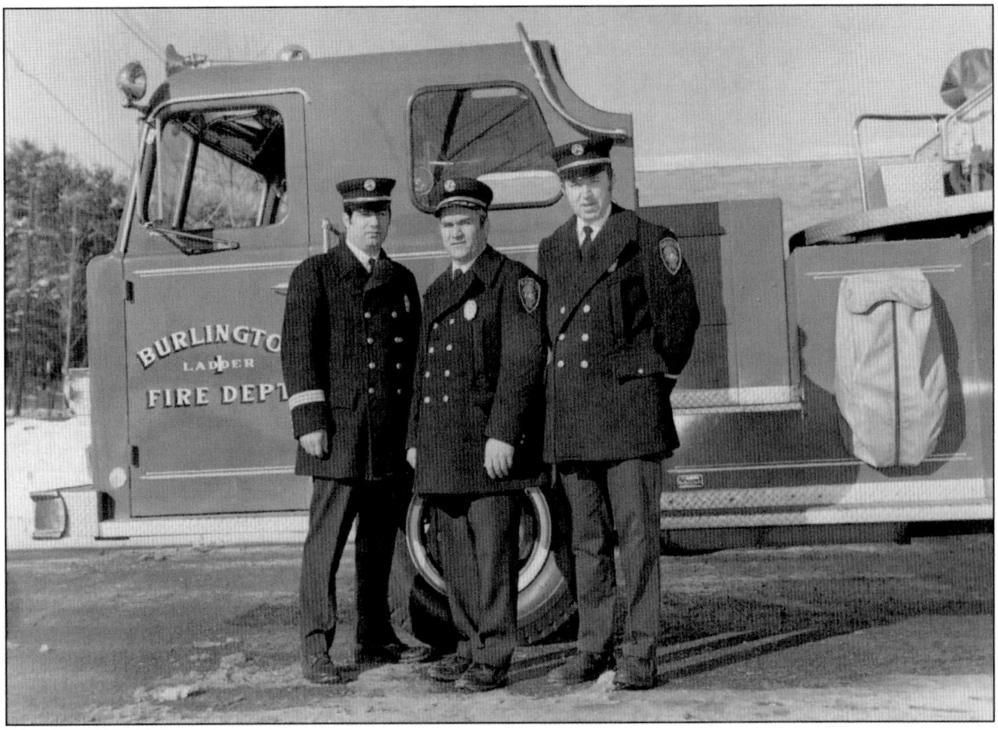

Acting chief Lawrence Rice (1984–1986) is the third person in this group of three. In this photograph are Phillip Pollicelli (left), Herbert Crawford, and Lawrence Rice. As chief, Lawrence Rice brought his experience and knowledge to the position and was dedicated to the profession of firefighting. (Courtesy of Lawrence Rice.)

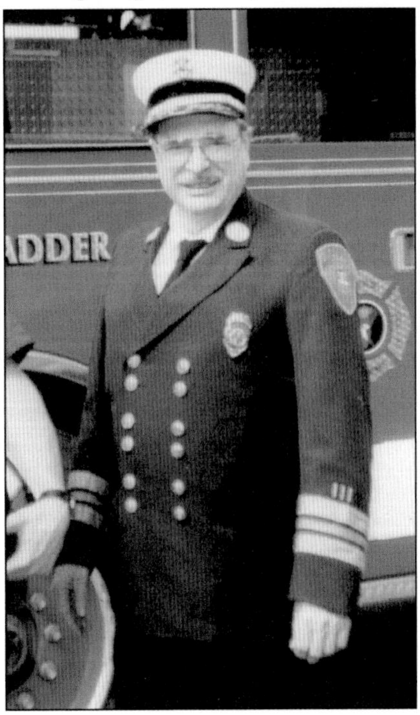

Chief Paul Thibault (1987–2004) was dedicated to his position as fire chief. He introduced new methods of firefighting and the use of computers to the department. He strived to improve the department and was untiring in his efforts to keep the department up to date. (Courtesy of Paul Thibault.)

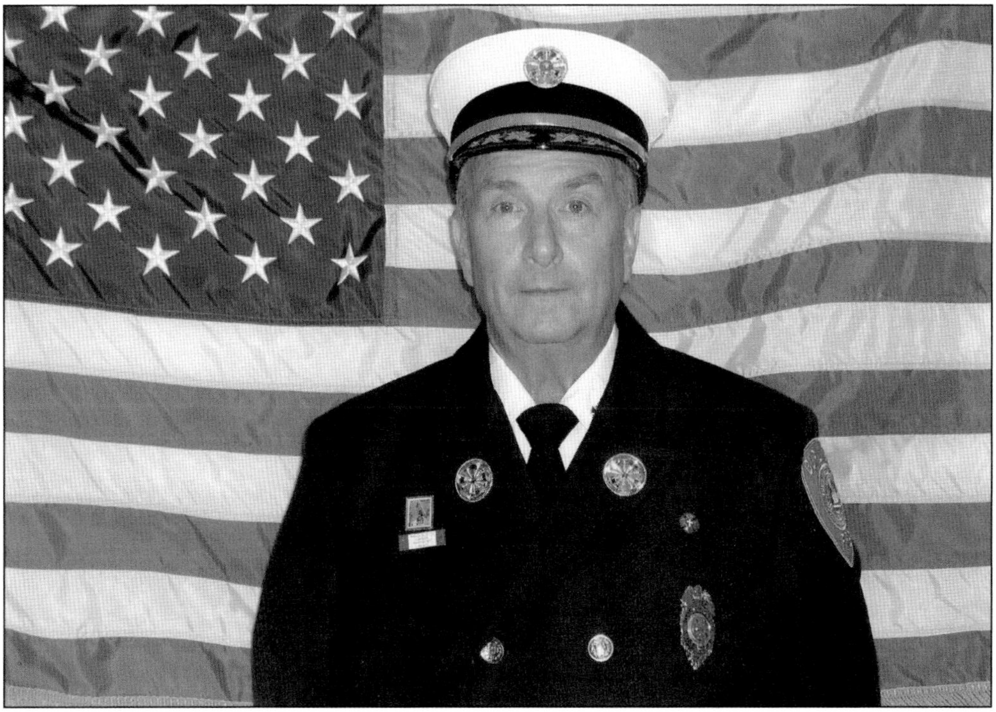

In keeping with present-day concerns, Chief Lee Callahan has the same challenges that the first fire chief, Herbert Crawford, faced. He too must keep abreast of the constant growth and varied changes in Burlington, but on a much larger scale. Today the face of Burlington is changing rapidly. Chief Callahan is encouraging Burlington firefighters to meet the challenges. They have responded by following through with the many training programs offered by the training officers. (Courtesy of Lee Callahan.)

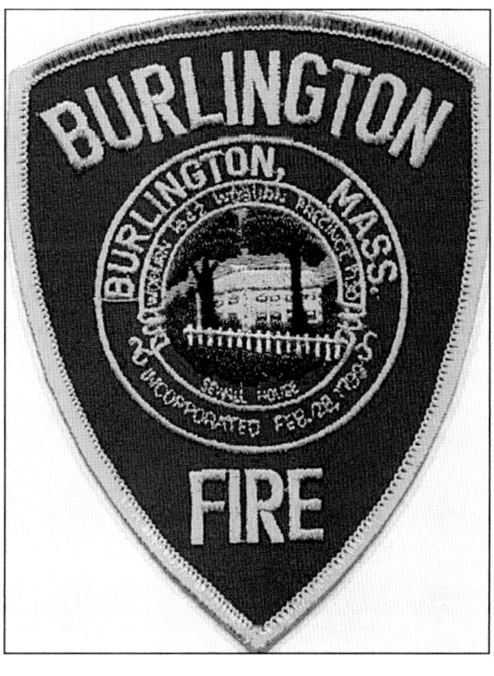

Pictured here is a Burlington fire patch. (Courtesy of Burlington Fire Department.)

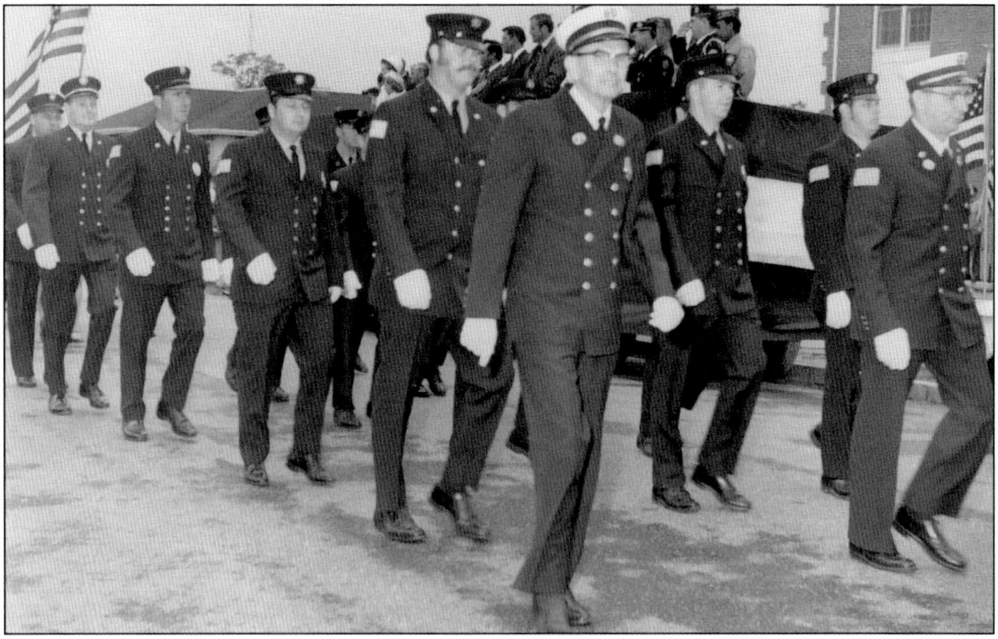

Officers Kenneth Brown and Louis Skelton march with their men past the town hall reviewing stand. From left to right are unidentified, unidentified, Kenneth Mills, Joseph Butler, Douglas Hyde, Kenneth Brown, Lawrence Rice, unidentified, and Louis Skelton. The date is unknown. (Courtesy of Albert Fay.)

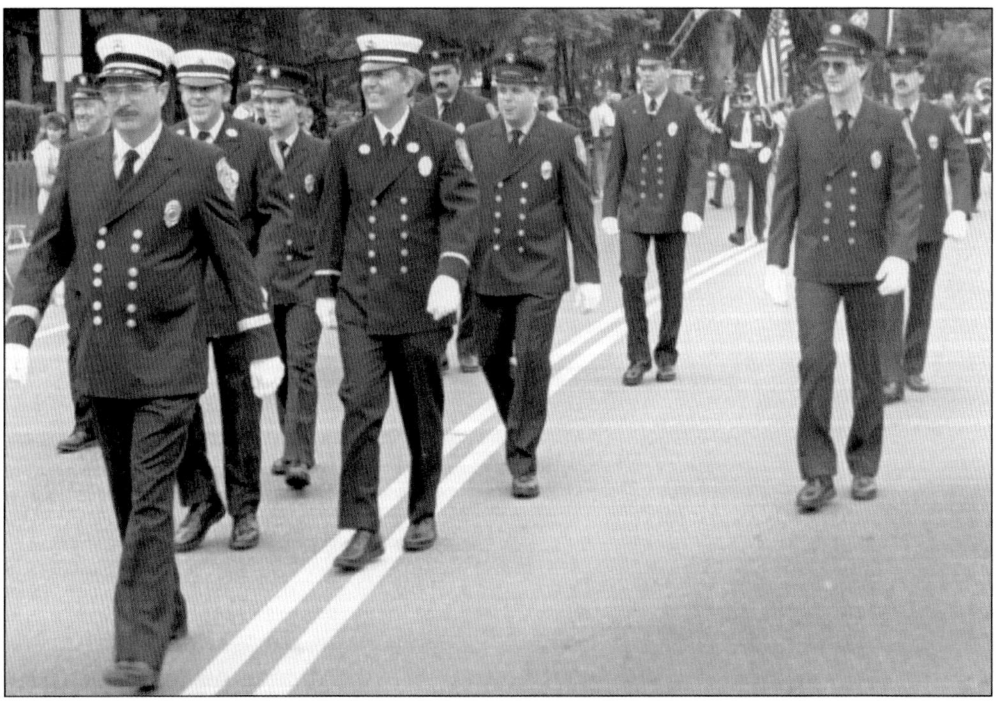

Chief Paul Thibault leads firefighters in a Memorial Day parade. From left to right are Frederick Keene, Chief Paul Thibault, Leo Pouliot, unidentified, William Devereaux, Albert Brabant, unidentified, unidentified, Scott Carpenter, and Paul Sullivan. (Courtesy of Albert Fay.)

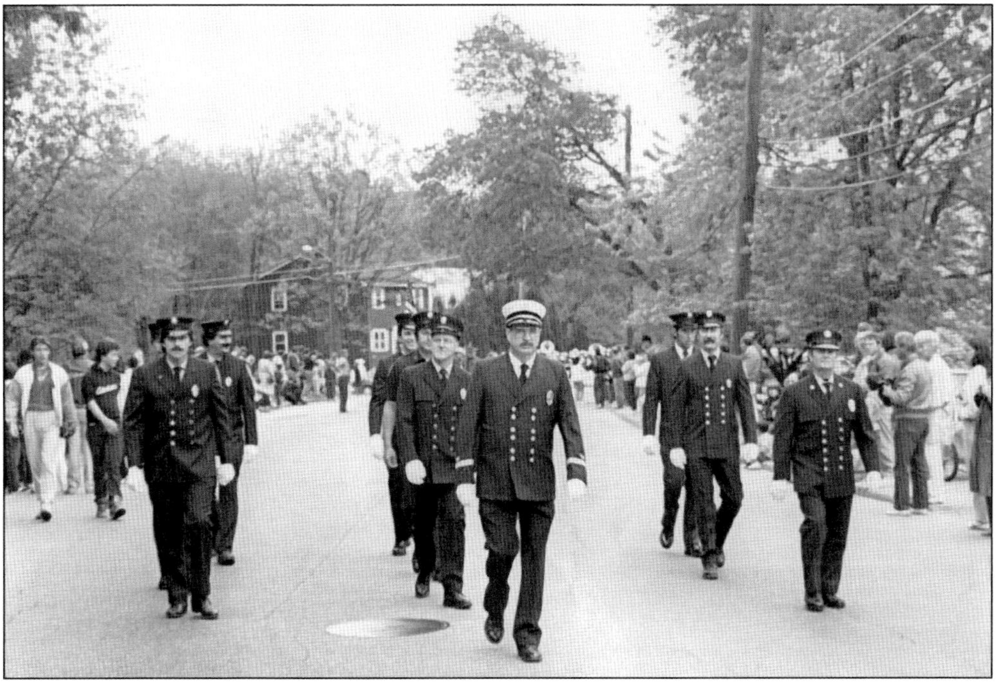

Chief Paul Thibault leads his men at an annual Memorial Day parade. Paul Sullivan is on the left. (Courtesy of Paul Thibault.)

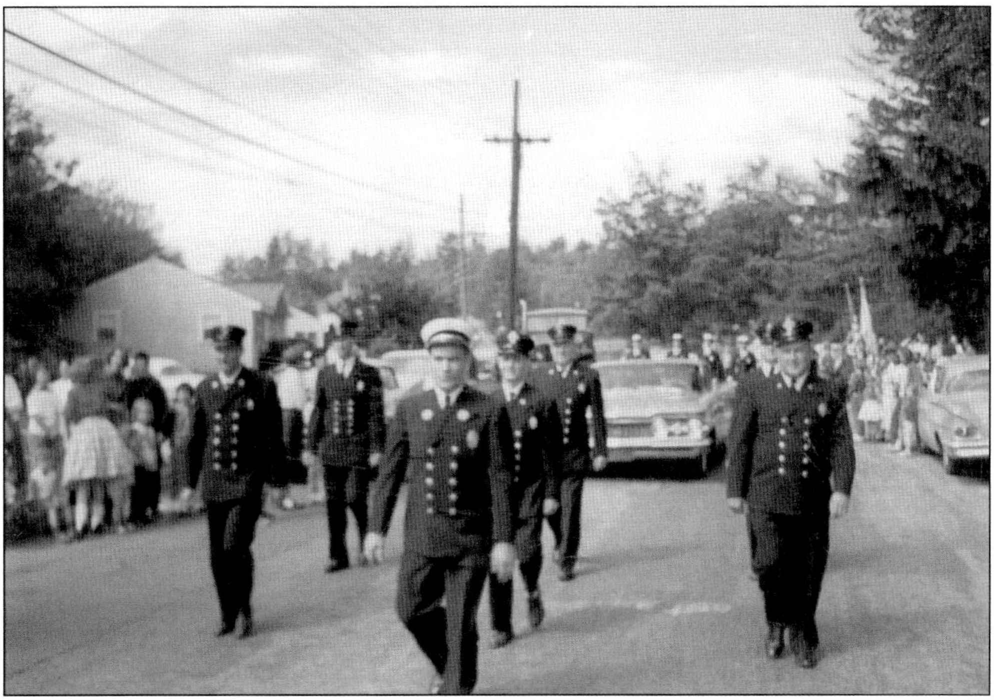

Chief Herbert Crawford and his department are shown marching in a Memorial Day parade. From left to right are unidentified, Lawrence Faria, Chief Crawford, Albert Fay, unidentified, Frederick Keene, and unidentified. The date is unknown. (Courtesy of Herbert Crawford.)

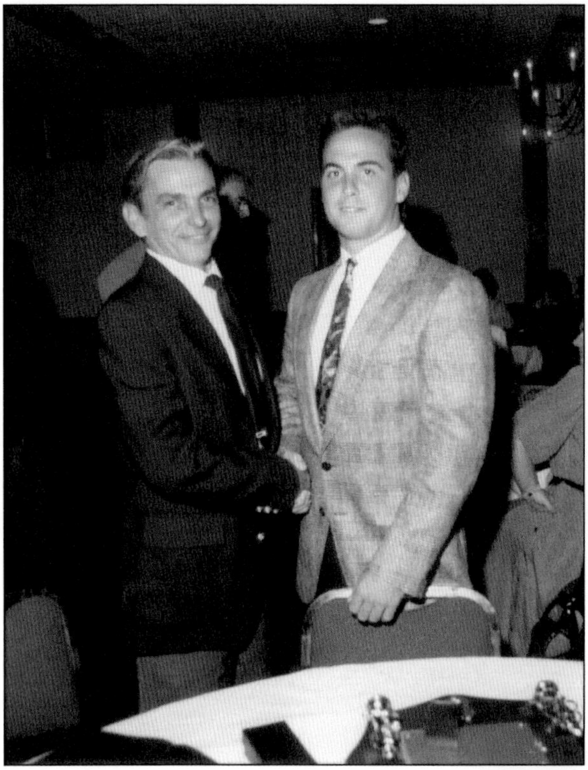

Fathers and sons are well represented on the department. William Bibbey was appointed a call firefighter on June 2, 1968. He was appointed a permanent firefighter on April 5, 1970, and retired on June 1, 1991. His son Michael Bibbey was appointed on September 26, 1988. (Courtesy of Michael Bibbey.)

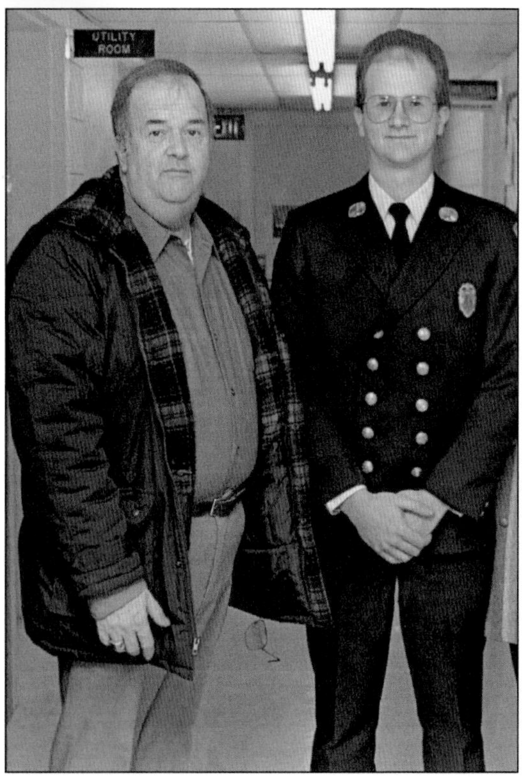

Capt. Scott Carpenter was appointed on February 7, 1985. He is the son of Robert Carpenter Jr., a call firefighter in the 1950s, and grandson of Robert Carpenter, a call firefighter in the 1930s. In addition, his sister Elaine Carpenter is a dispatcher on the department. (Courtesy of Scott Carpenter.)

William Callahan was appointed on April 30, 1978, and retired on April 28, 2007. His son Craig Callahan is planning to follow his father's example and is soon to be appointed. (Courtesy of William Callahan.)

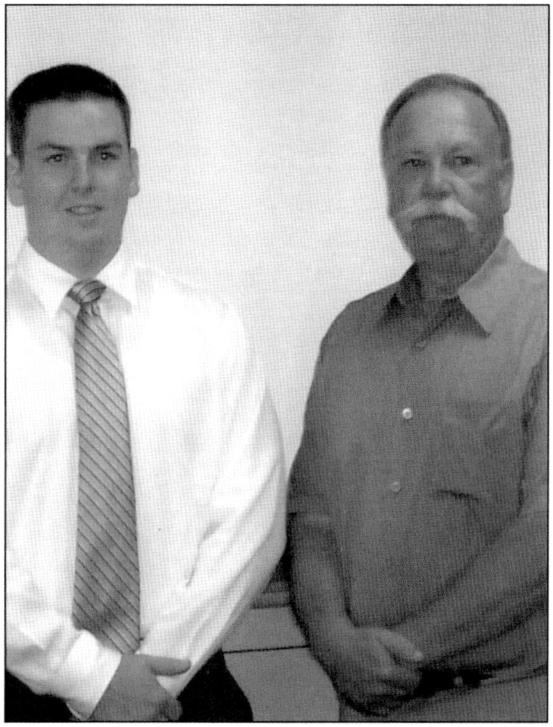

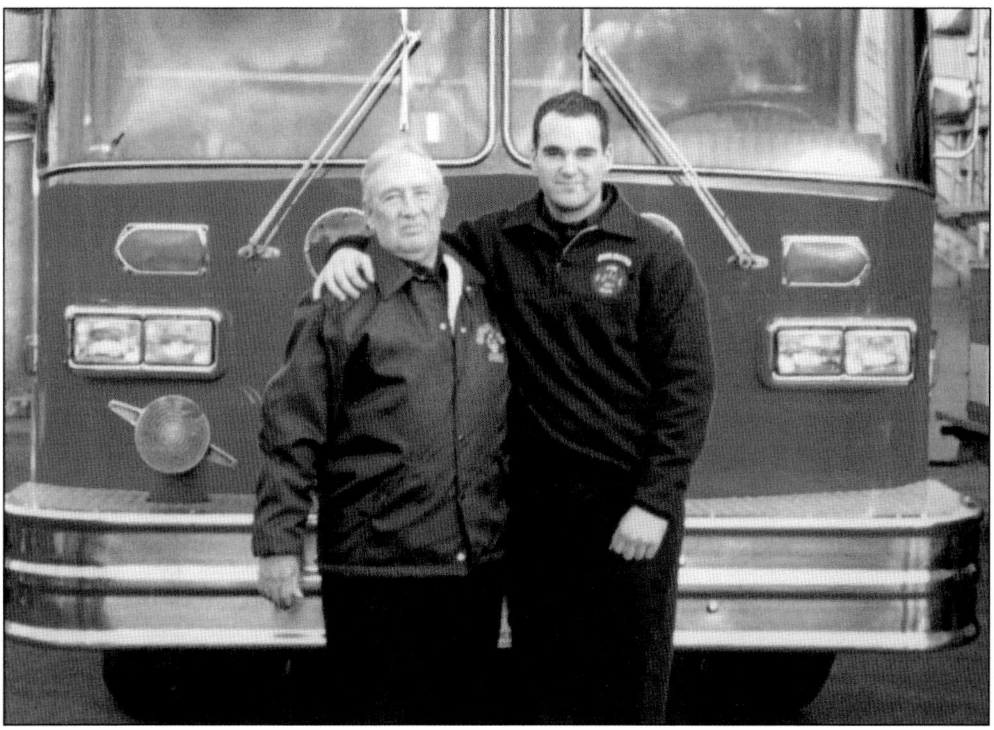

James Hapenney Sr. was appointed on November 25, 1973, and retired in February 2000. His son James Hapenney Jr. was appointed on September 13, 1999, which allowed them to be on the department together for a short time. (Courtesy of James Hapenney Sr.)

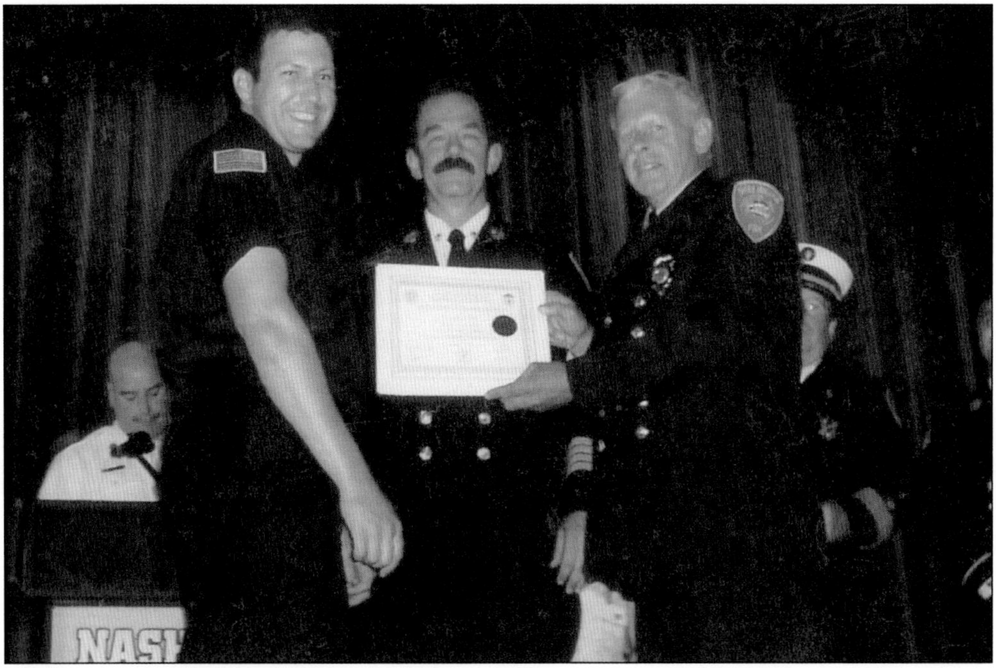

Paul McLaughlin was appointed on April 13, 1975, and retired on November 15, 2000. His son Michael McLaughlin was appointed on July 25, 2005. From left to right are Michael McLaughlin, Steven Yetman, and Paul McLaughlin. (Courtesy of Paul McLaughlin.)

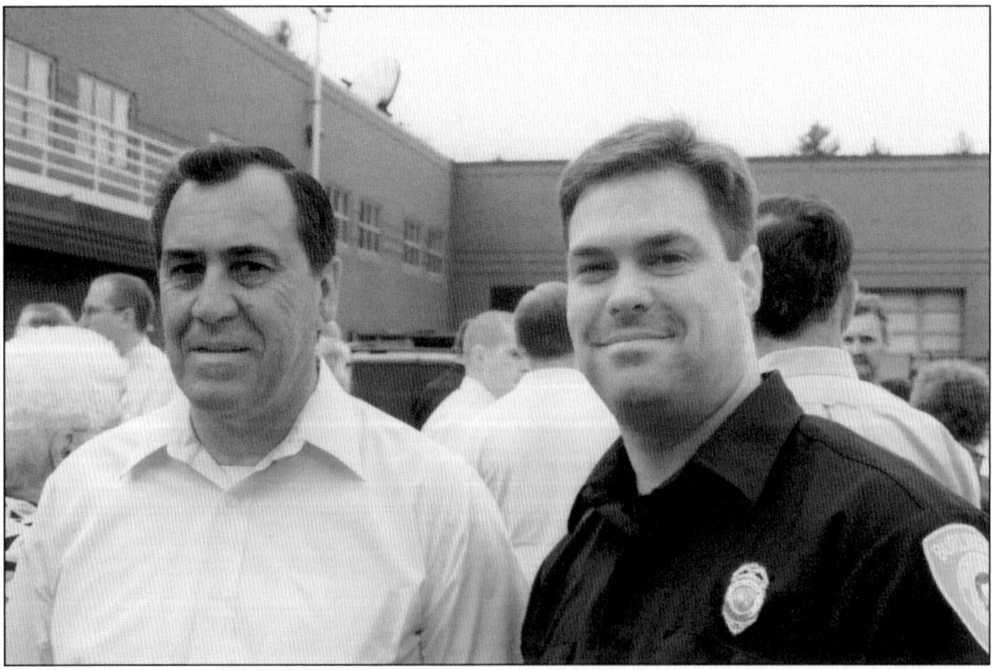

Raymond Perry was appointed a call firefighter on June 23, 1970, and a permanent firefighter on September 21, 1970. He retired on December 26, 1998. His son Scott Perry was appointed on February 20, 2001, and served until April 3, 2006. He is currently a dispatcher on the department. (Courtesy of Scott Perry.)

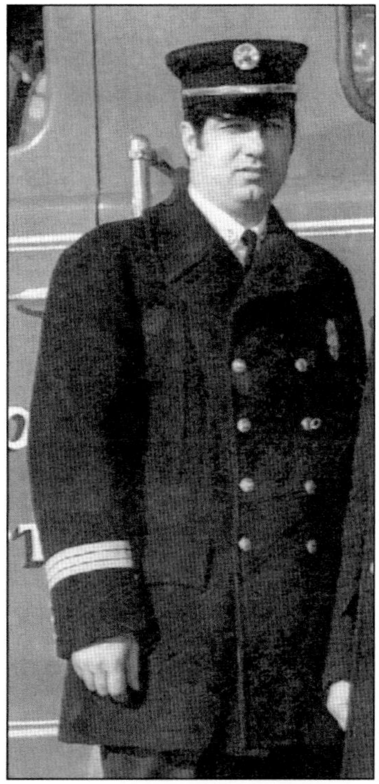

Phillip Pollicelli was appointed on January 4, 1964, and served until December 24, 1992. His son Kevin Pollicelli was appointed on September 13, 1999. He is shown in the photograph below with Chief Paul Thibault. (Right, courtesy of Burlington Fire Department; below, courtesy of Kevin Pollicelli.)

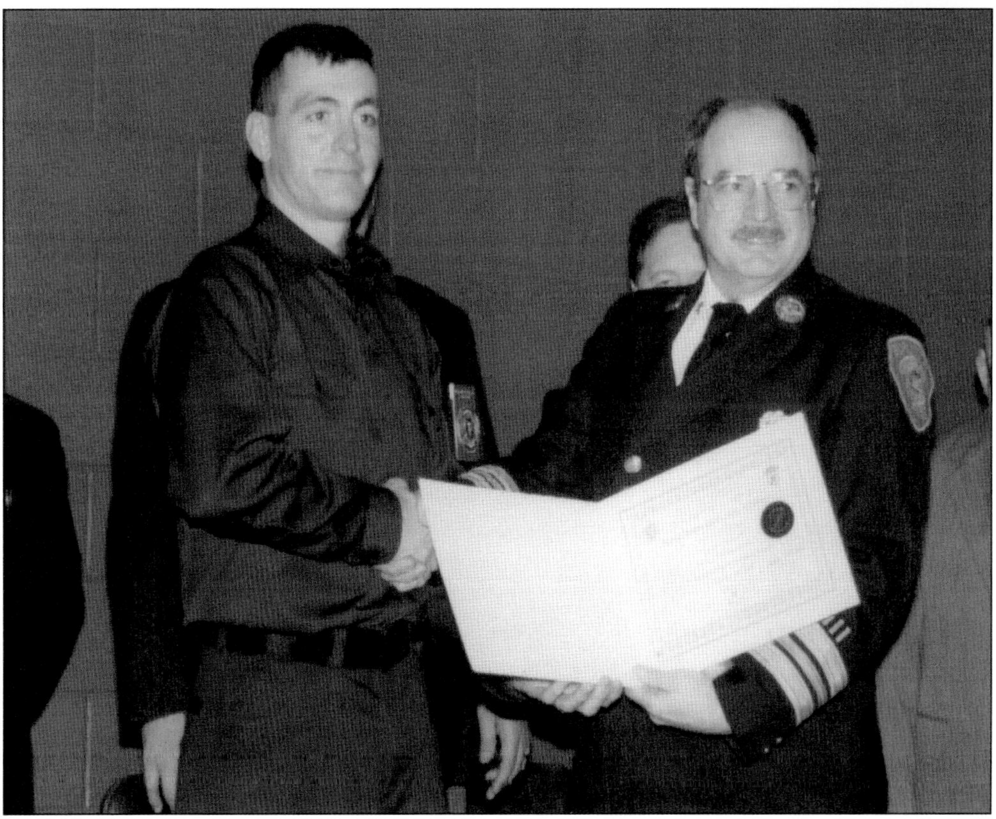

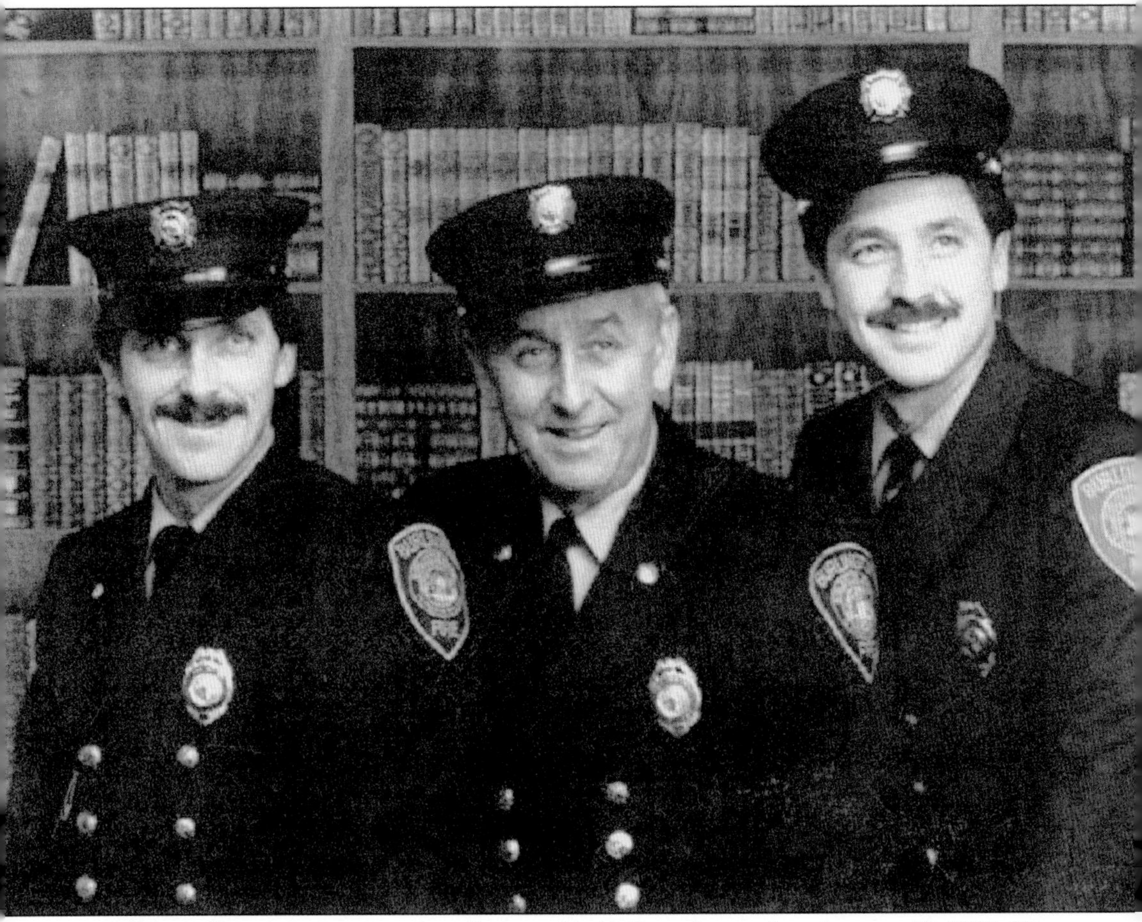

Robert Sullivan was appointed on January 14, 1962, and retired on January 15, 1988. His son Paul (left) was appointed on July 1, 1979, and son David was appointed on November 3, 1985. (Courtesy of Robert Sullivan.)

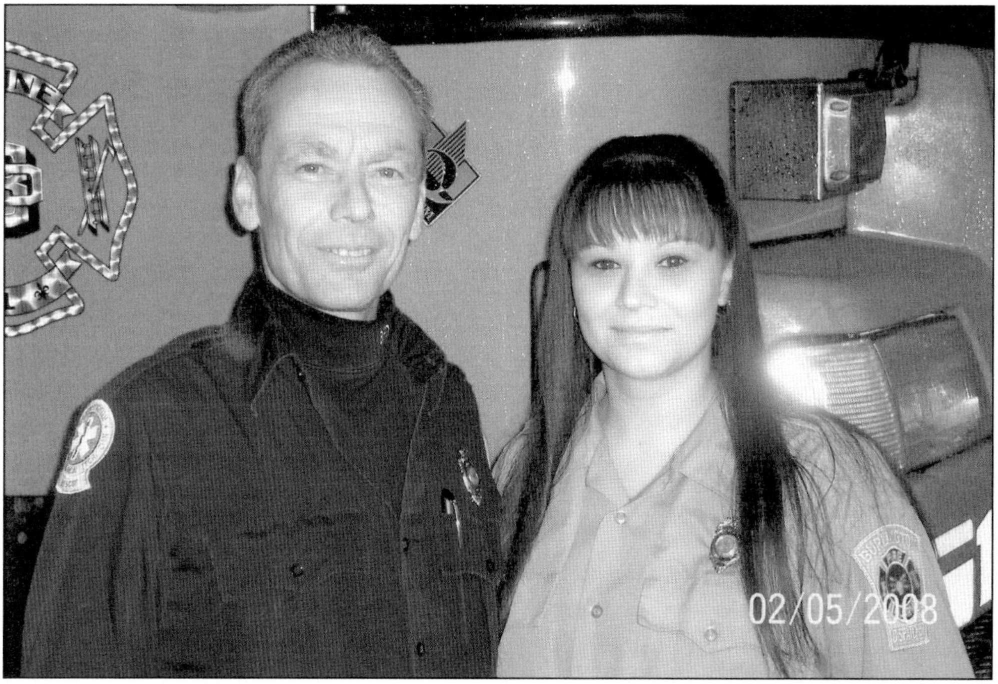

Firefighter Michael Runyan is joined on the department by his daughter Nicole Runyan DeTomasi, a dispatcher. (Courtesy of Nicole DeTomasi.)

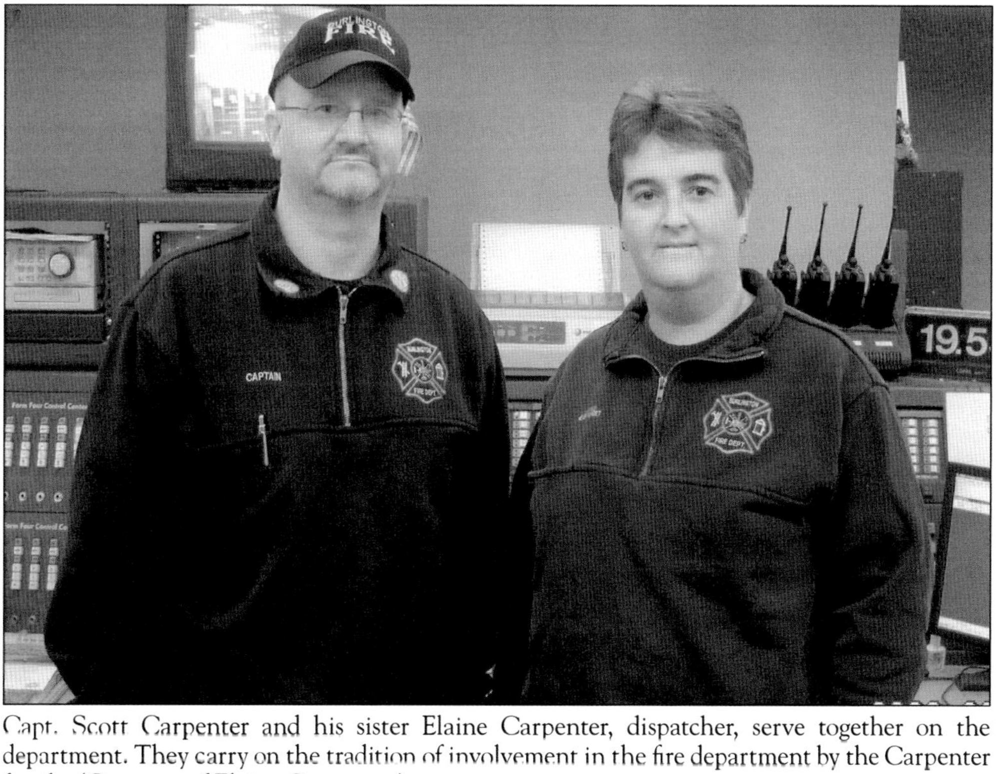

Capt. Scott Carpenter and his sister Elaine Carpenter, dispatcher, serve together on the department. They carry on the tradition of involvement in the fire department by the Carpenter family. (Courtesy of Elaine Carpenter.)

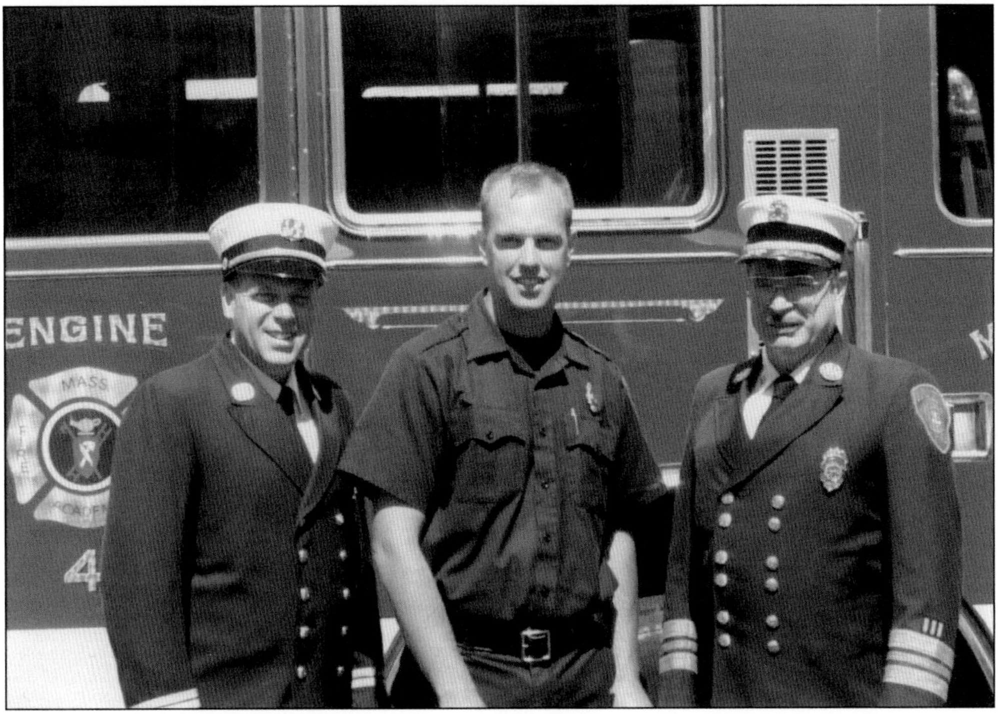

Brothers Lt. James Browne (center) and Capt. Kevin Browne (left) are shown here with Fire Chief Paul Thibault. Not in the photograph is another brother, Lt. Timothy Browne. They are joined by four sets of brothers who are also firefighters. They are John and Michael Hanafin, Richard and Timothy Hovasse, and David and Paul Sullivan, as well as two retired firefighters, Albert and James Fay. Brothers William Callahan and Chief Lee Callahan served on the department together before William retired in 2007. (Courtesy of Paul Thibault.)

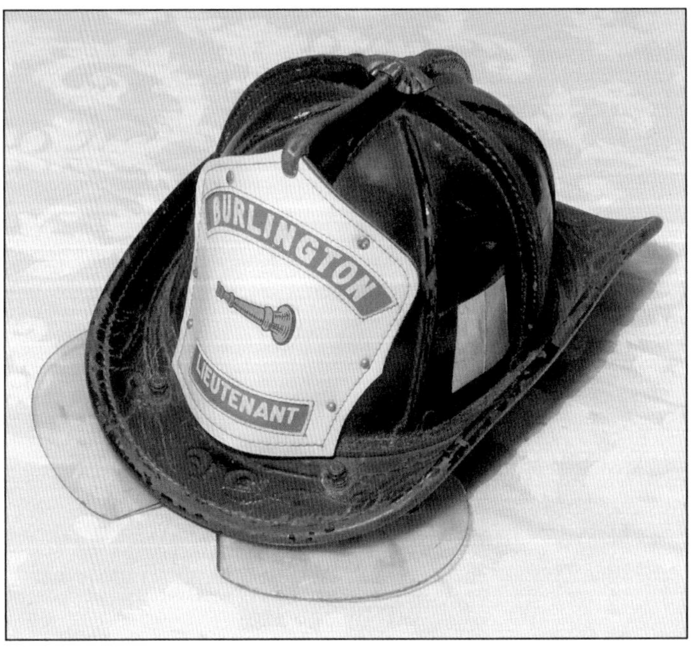

A helmet, the international symbol of a firefighter, is a vital part of a firefighter's equipment. It has retained the same look for generations. (Courtesy of Michael Tredeau.)

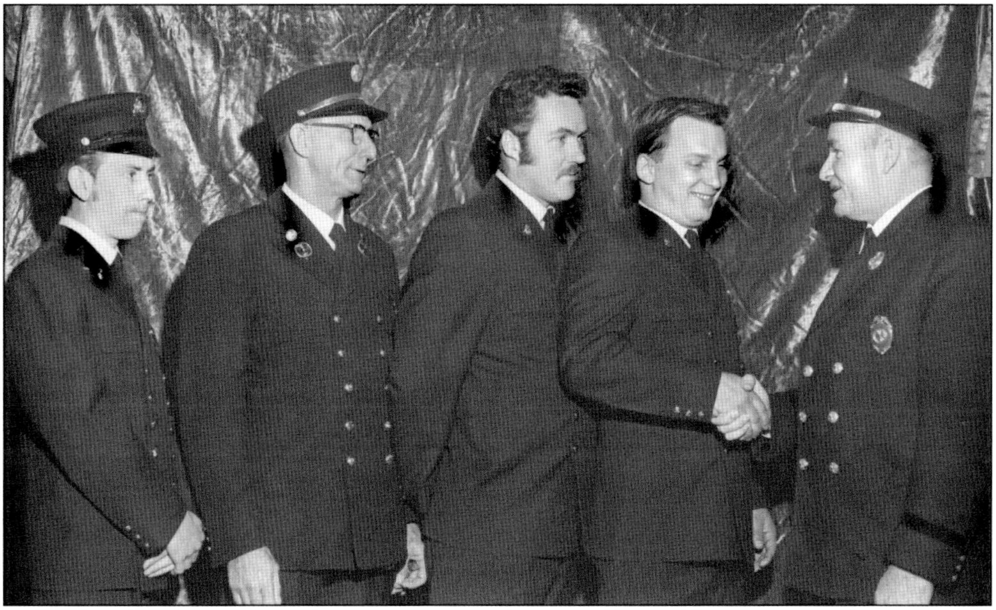

Events past and present include a graduation from the fire academy in 1969. In attendance was Louis Skelton, second from the left, and congratulating the men (from left to right, James Fay, Joseph Doran, and Donald Robinson) is Chief Herbert Crawford. (Courtesy of Albert Fay.)

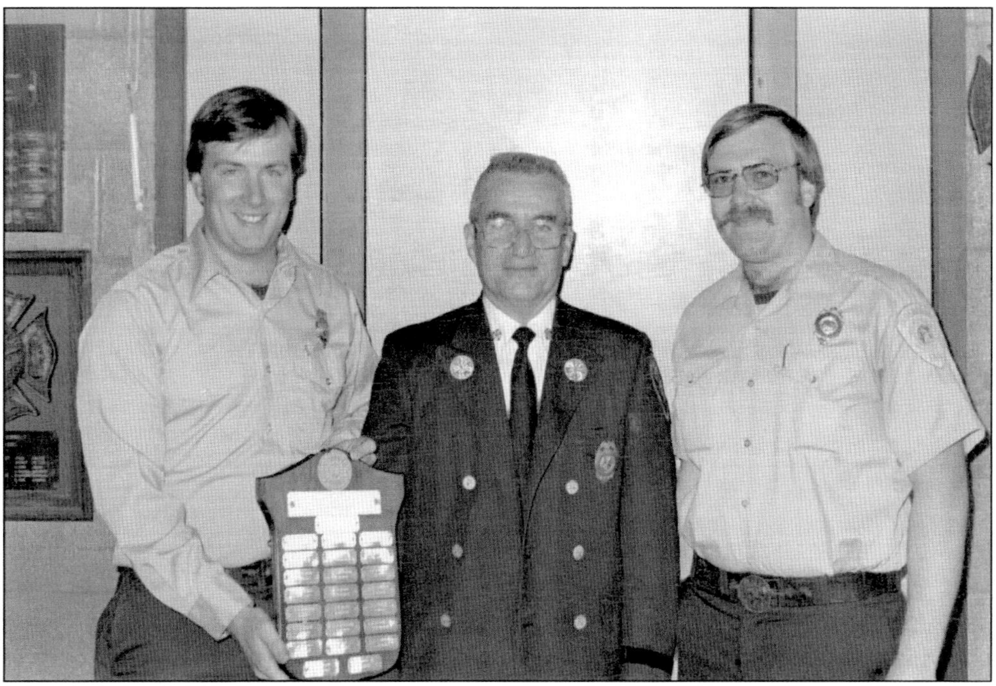

At the graduation from the fire academy in 1978, Lee Callahan, Fire Chief Herbert Crawford, and William Callahan, from left to right, stand together. Brothers Lee and William Callahan are the nephews of Chief Herbert Crawford. (Courtesy of Lee Callahan.)

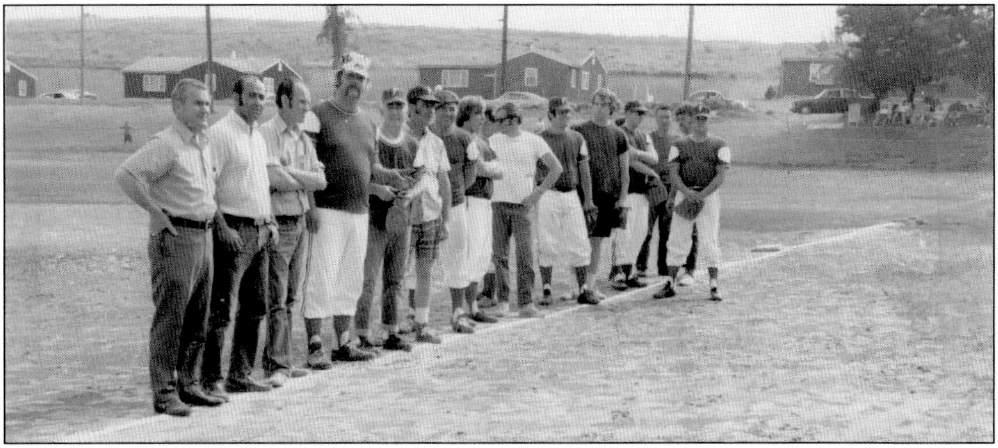

Pictured here is a fire department softball game in 1971. From left to right are Chief Herbert Crawford, Robert Guidoboni, Robert Ferro, Douglas Hyde, Albert Fay, unidentified, unidentified, Daniel Willete, George Clark, and Phillip Pollicelli; the remaining men are unidentified. (Courtesy of Herbert Crawford.)

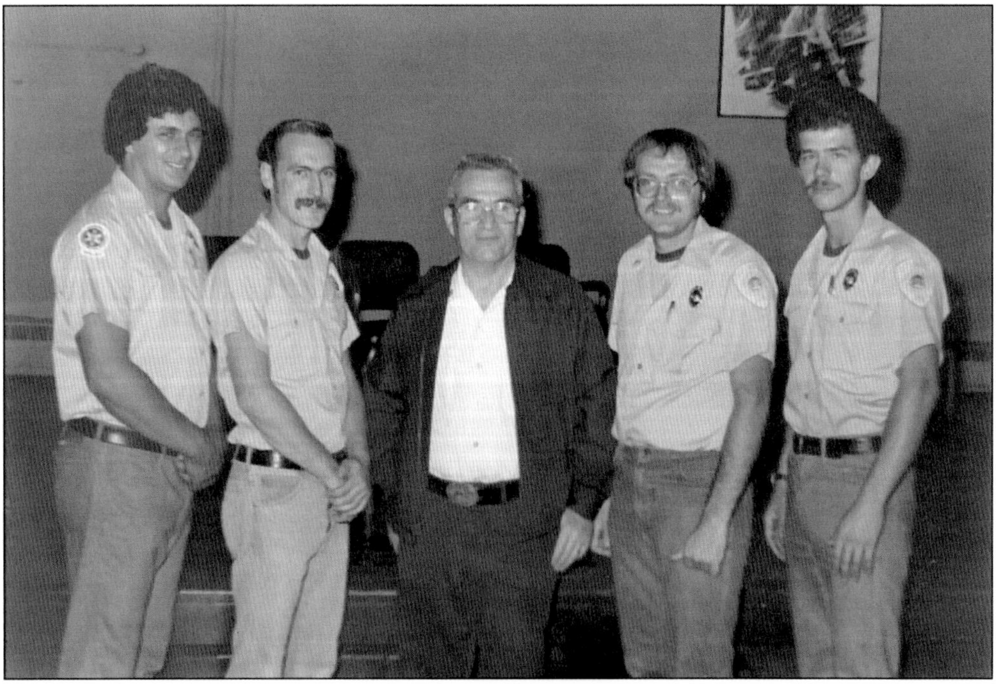

From left to right, Albert Brabant, Gary Cauldwell, Chief Herbert Crawford, William Gangestad, and Steven Yetman are seen here in 1978. (Courtesy of Albert Fay.)

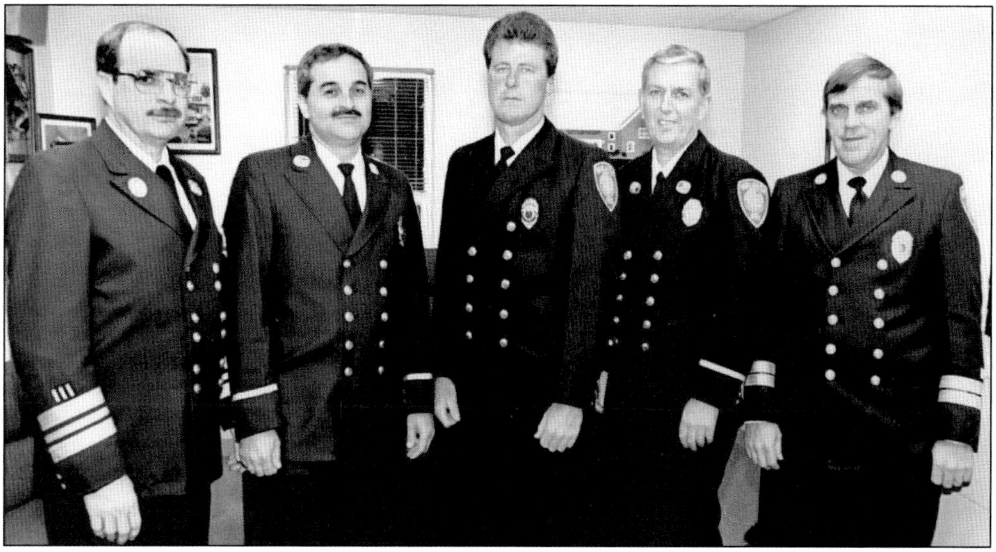

From left to right, Fire Chief Paul Thibault, Edward Costa, Michael Nolan, William Devereaux, and Leo Pouliot are in full uniform. (Courtesy of Albert Fay.)

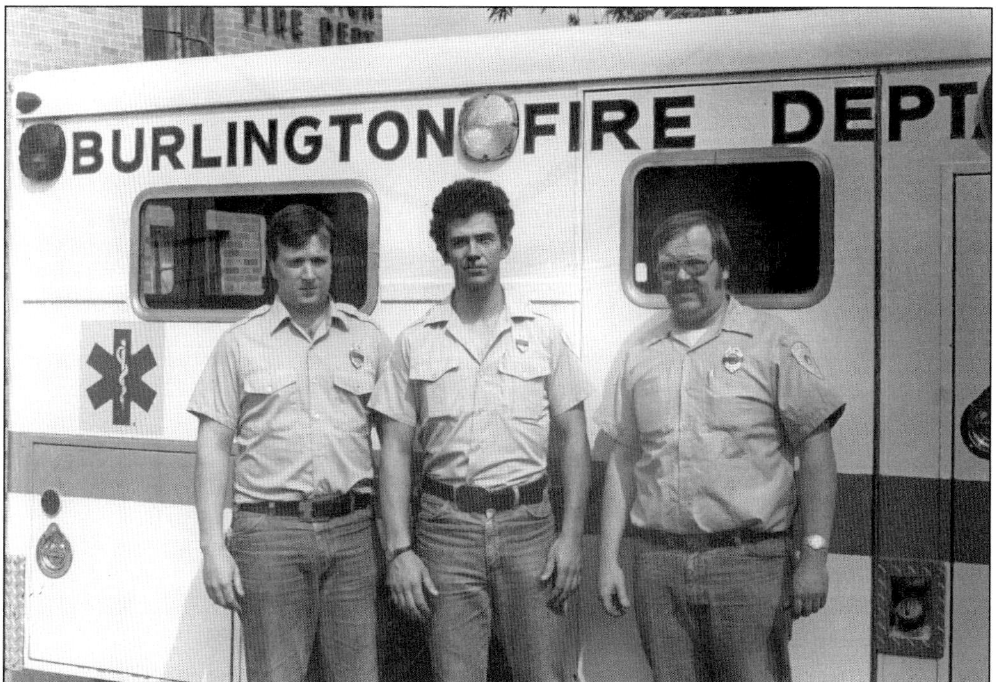

This 1981 photograph shows, from left to right, Lee Callahan, Steven Yetman, and William Callahan. Lee became fire chief in 2004, and Steven became the assistant chief in 2004. Lee and William Callahan are brothers. (Courtesy of Lee Callahan.)

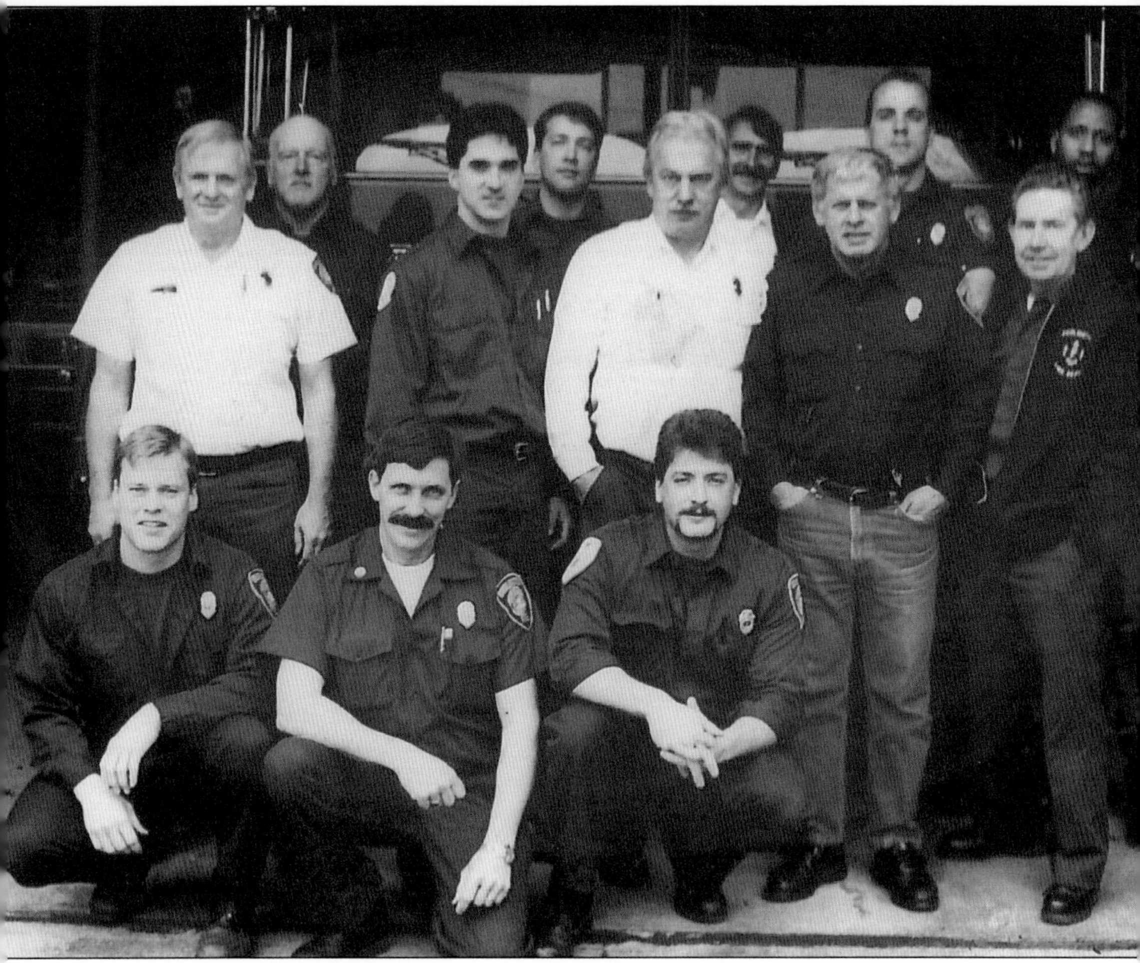

Group three is a representation of older and younger men who worked well together. They are, from left to right, (first row) Robert Paul, Paul Sullivan, and Richard Hovasse; (second row) George Clark, Michael Hanafin, Donald Robinson, Paul McLaughlin, and James Fay; (third row) Frederick Keene, Kurt Duprez, George Creamer, Gerald Hanafin, and Frederick Williams. (Courtesy of Susan Clark.)

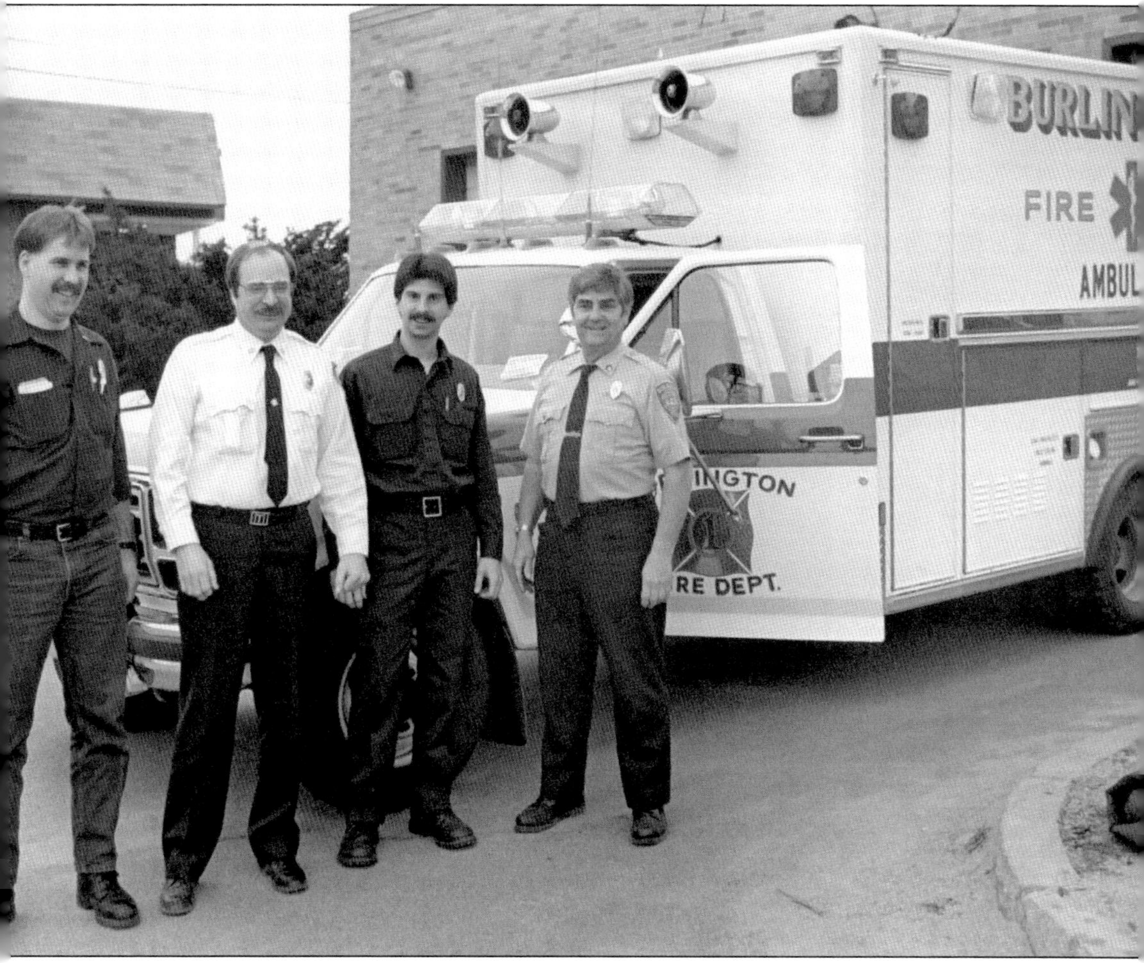

From left to right, Gary Arbing, Chief Paul Thibault, unidentified, and Albert Fay stand in front of an ambulance at headquarters. (Courtesy of Albert Fay.)

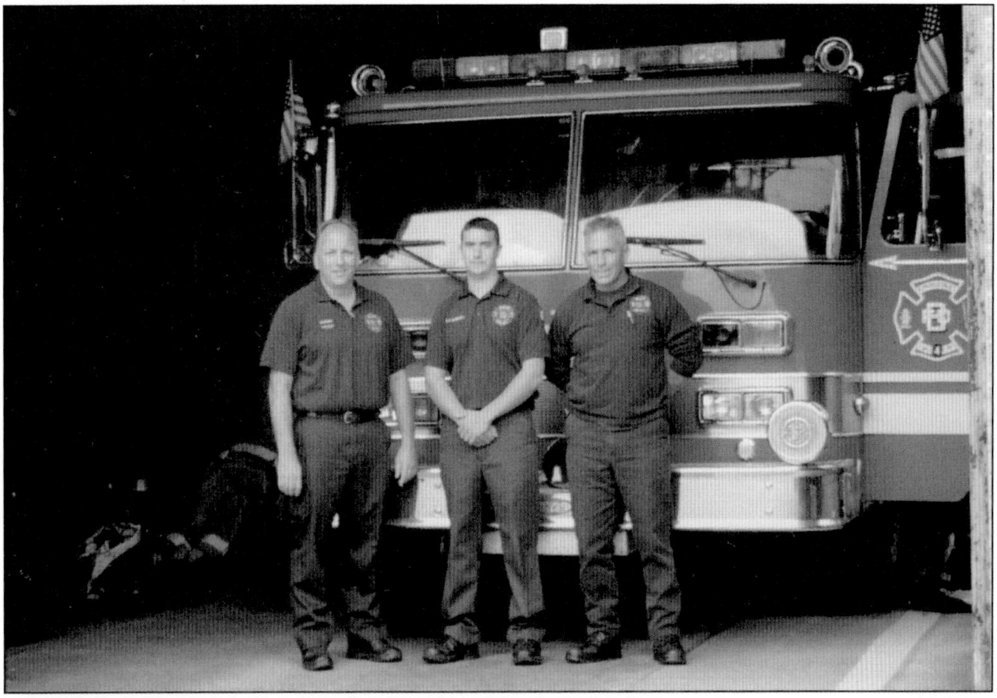

From left to right, Peter McAnespie, Michael Glendhill, and James Sherman are shown in this recent photograph taken at the substation on Terrace Hall Avenue. (Courtesy of Toni Faria.)

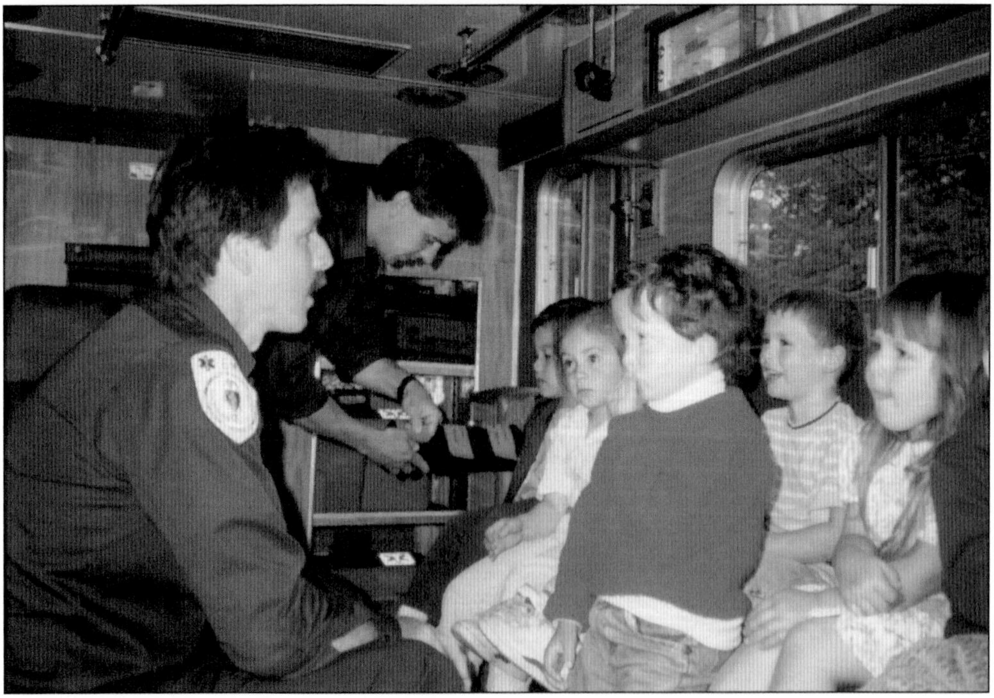

David Sullivan (left) and James Sorenson teach fire safety to children at headquarters. Field trips from elementary schools are an opportunity to teach children about fire prevention. (Courtesy of Albert Fay.)

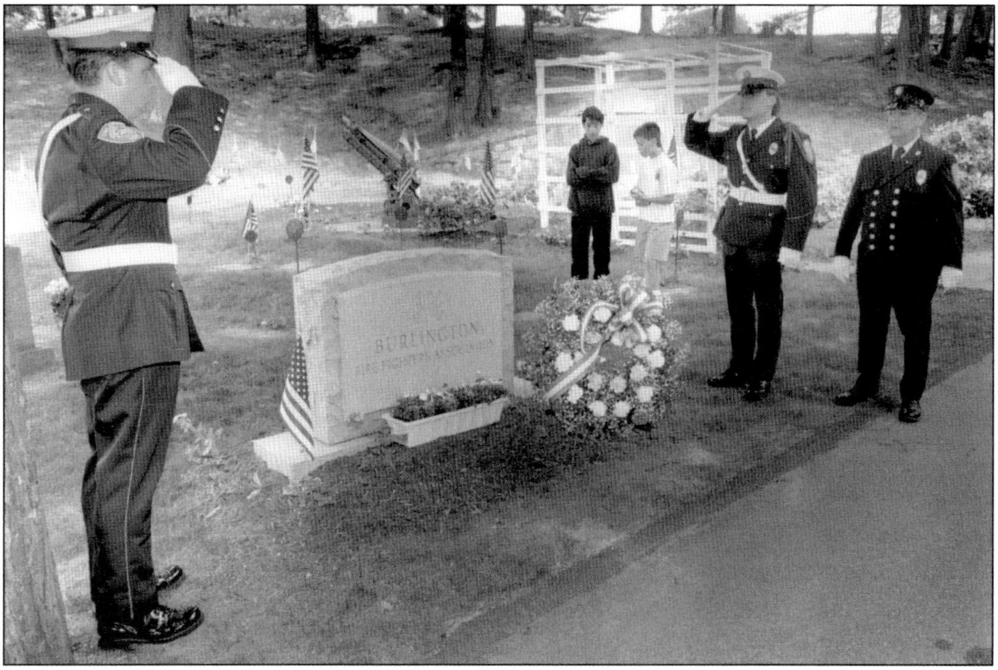

From left to right, Burlington firefighters Gerald Letendre, Robert Paul, and Albert Fay salute the fallen firefighters in Burlington. (Courtesy of Albert Fay, photograph by Joe Brown.)

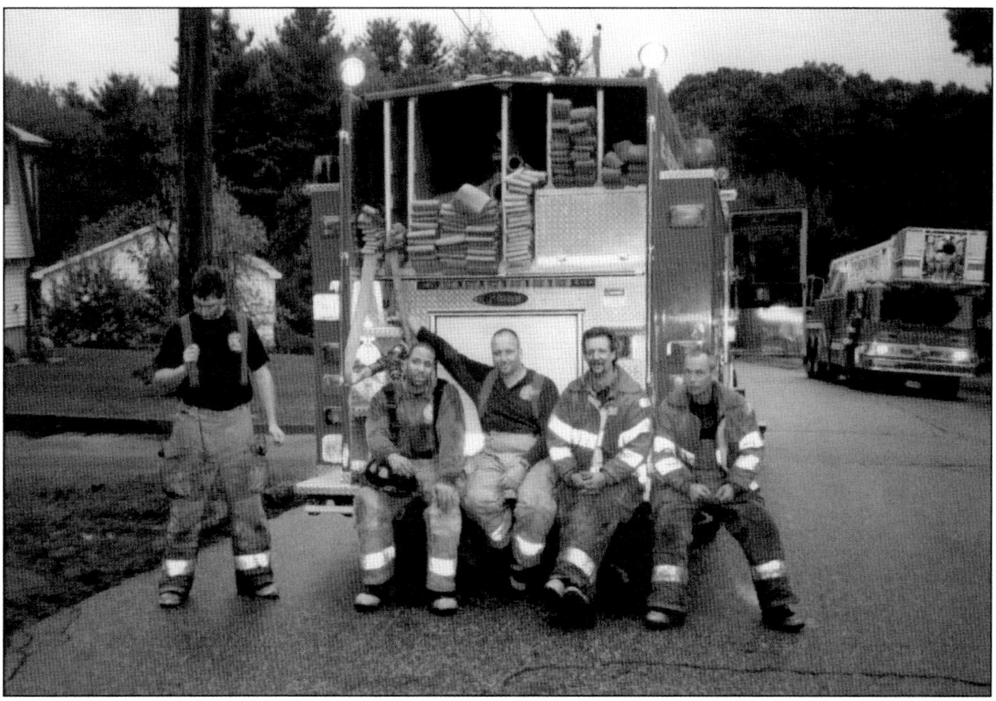

From left to right, firefighters Richard Hovasse, Frederick Williams, Peter McAnespie, Michael Donahue, and Michael Runyan are seated at the back of the equipment. (Courtesy of Burlington Fire Department.)

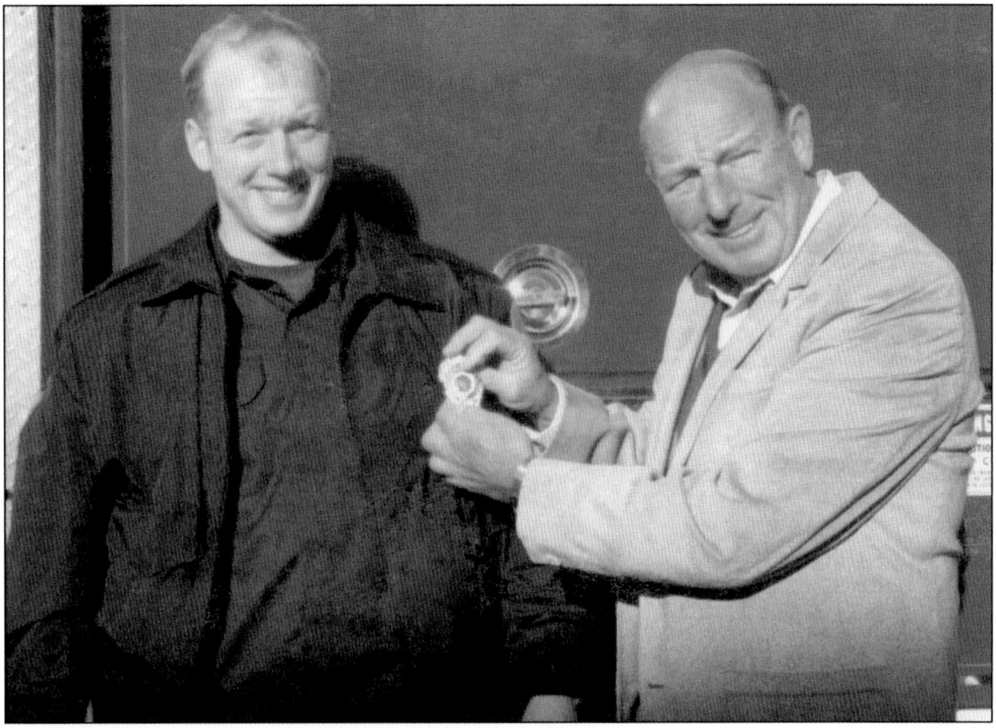

In 1998, Sean Connors and his father, local resident Michael Connors, are shown here as Sean's badge is pinned on by his proud father. (Courtesy of Paul Thibault.)

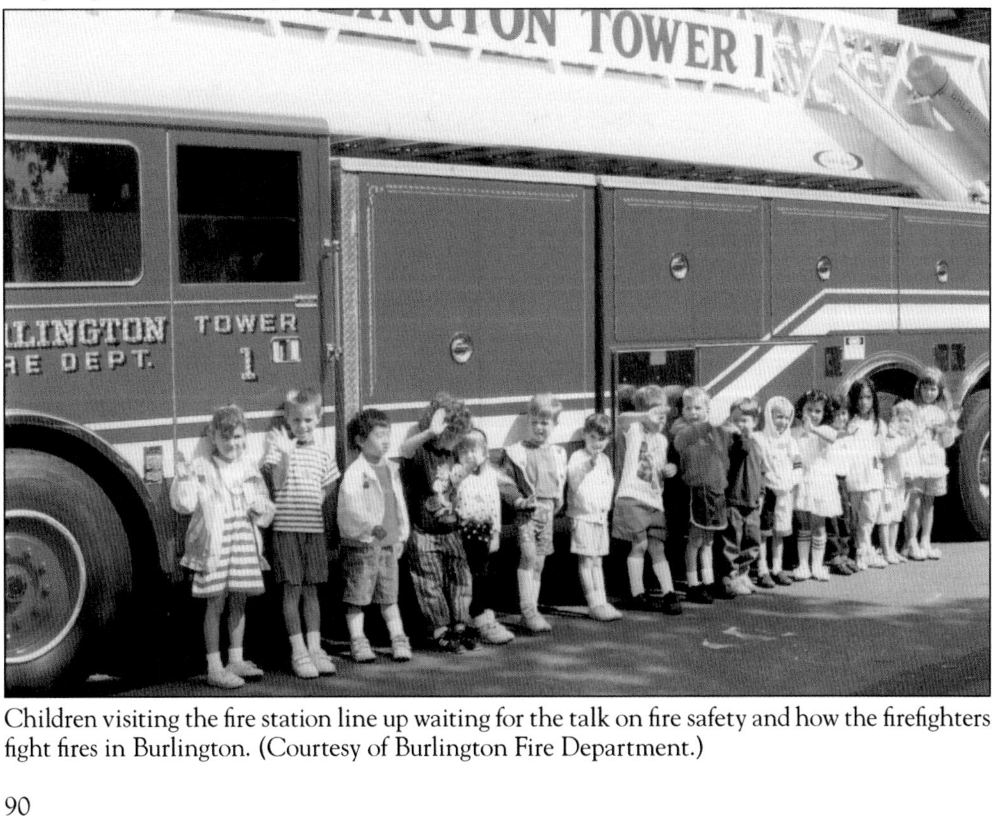

Children visiting the fire station line up waiting for the talk on fire safety and how the firefighters fight fires in Burlington. (Courtesy of Burlington Fire Department.)

Seven

The Skelton Family

The first Skeltons came to America in 1629 and settled in Salem. They lived in Woburn as early as 1653. In 1692, they were given a parcel of land on what is now Francis Wyman Road. Through the years they have been involved in community affairs and have fought in their country's wars from the Revolution to Vietnam. Unlike some early settlers, who no longer live in Burlington, they have remained residents of Burlington. They have been socially and politically active in the community. The Skelton family has been leaders in volunteer firefighting since it began. Walter, Orray, Lester, and Horace were all members of the group.

As the fire warden in 1932, Walter was instrumental in bringing a water system to Burlington. He was on the water investigating committee to petition the state legislature for permission to "supply itself and its inhabitants with water for the extinguishment of fires and for domestic and other purposes." Several years went by before the results of those actions had any effect.

Walter Skelton was the first fire chief in 1935. Louis Skelton served on the fire department from 1955 to 1978. The Burlington Fire Department dedicated the lobby of the headquarters in honor of the Skeltons' contribution to firefighting.

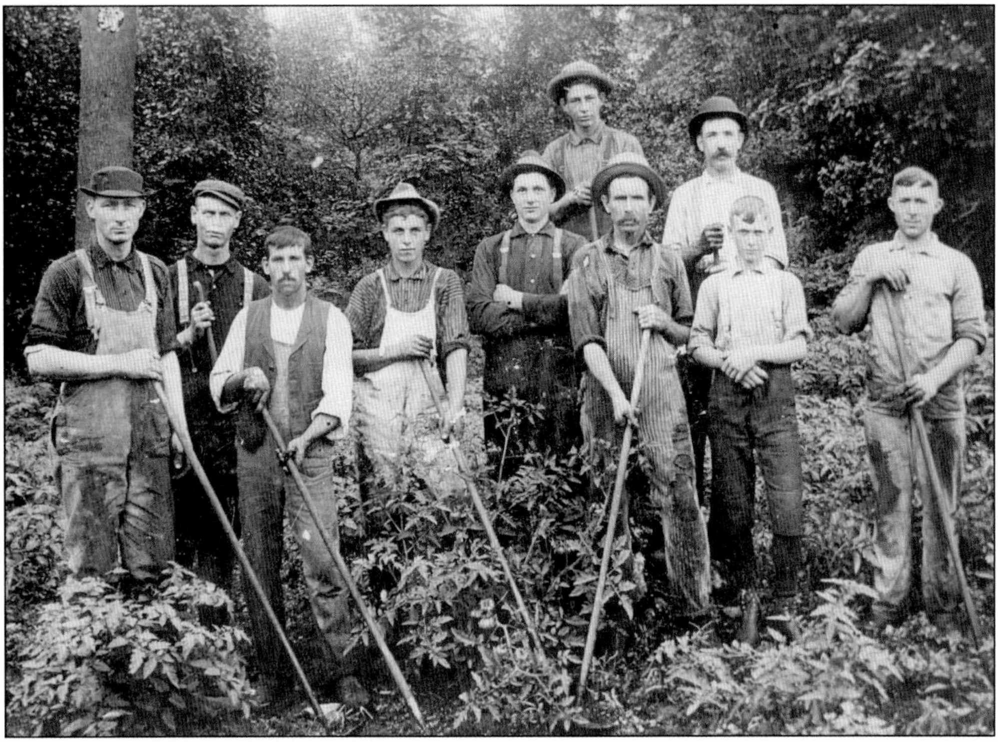

The Skelton family has been well represented in every century since arriving in 1662. In this early photograph of farmers, Orray Skelton, on the far left, is shown with farmhands. When the volunteers began organizing to fight fires, Orray was an active member. (Courtesy of Burlington archives.)

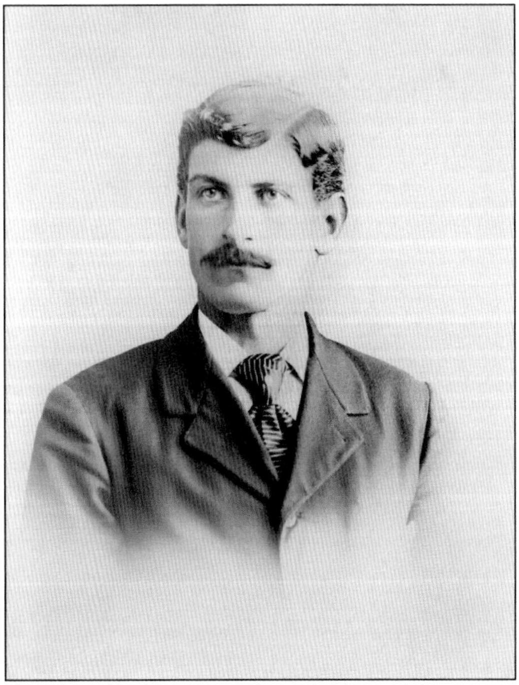

Born in 1864, Walter Skelton was the eldest of four boys born to Bradford and Almira Skelton. Lester, Orray, and Horace were the younger sons. Walter became interested in the importance of firefighting when he was an adult and became the leader of the volunteer firefighters. Walter served as selectman, assessor, school committee member, and cemetery commissioner. He was the fire chief from 1935 to 1937. (Courtesy of Burlington archives.)

Horace Bradford Skelton was born in 1879. He had a house and farm on Francis Wyman Road. The Skelton family became involved in firefighting when Burlington was still part of Woburn and firefighting was a volunteer effort. Along with his brothers, he was active in the early firefighting organization. (Courtesy of Burlington archives.)

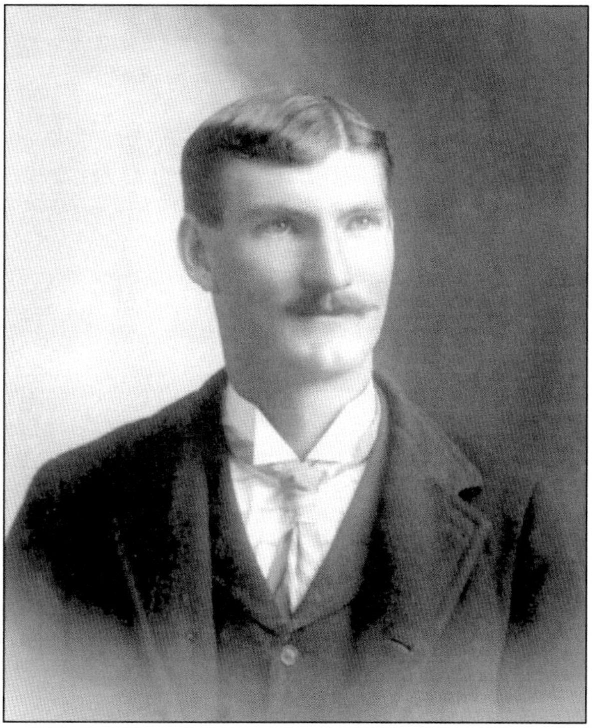

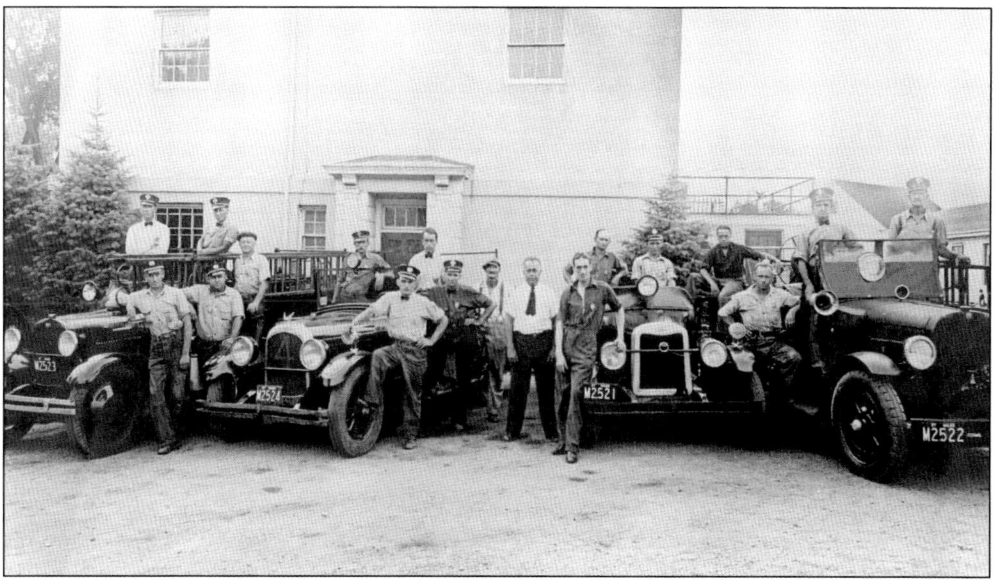

Members of the Skelton family are well represented once again, as they pose in this 1937 photograph. Four of the 18 men in this photograph, taken with four pieces of equipment, are from the Skelton family. They are David, Louis, Horace, and Marshall Skelton. Cliff Skelton was also a call firefighter but is not in the photograph. Pictured are, from left to right, (first row) Chet Knowles, Ted Parkhurst, David Skelton, Louis Skelton, Horace Skelton, George Martikke, Wilbur McIntire, and Marshall Skelton; (second row) Ray LeFave, Charles Bunton, David Lundin, Ralph Knowles, Sidney Brown, Bob Whitehouse, Chet Young, Dick Cronin, Edward Miocchi, and George Beard. (Courtesy of Burlington archives.)

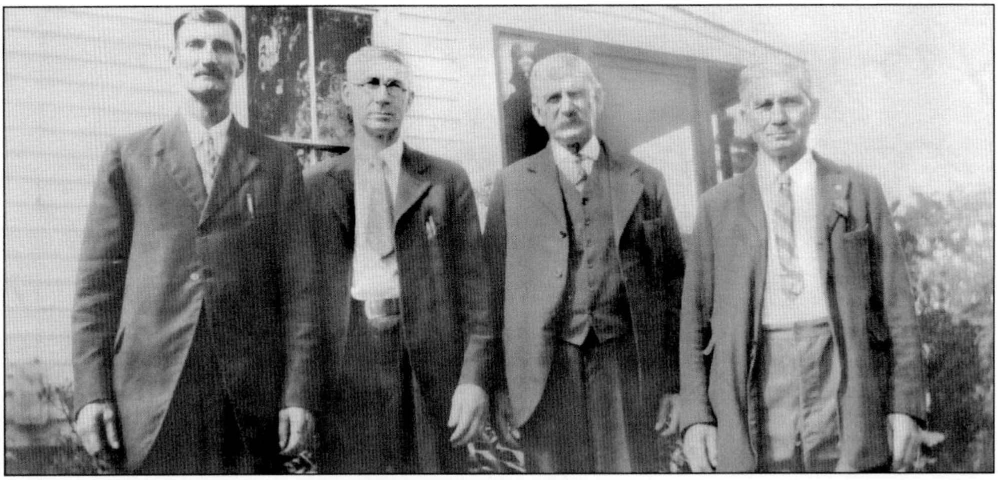

The four Skelton brothers, Lester, Walter, Orray, and Horace, are shown. All four brothers grew up on the farm on Francis Wyman Road. They grew up in Burlington, attended the Union School, and led productive lives in Burlington. Walter began his interest in the fire department and tried to have the town fathers act on his ideas for better fire protection. Burlington had no firefighting equipment until 1909, when the town authorized a wagon and 25 milk cans for use by the fire warden, Walter Skelton. (Courtesy of Burlington archives.)

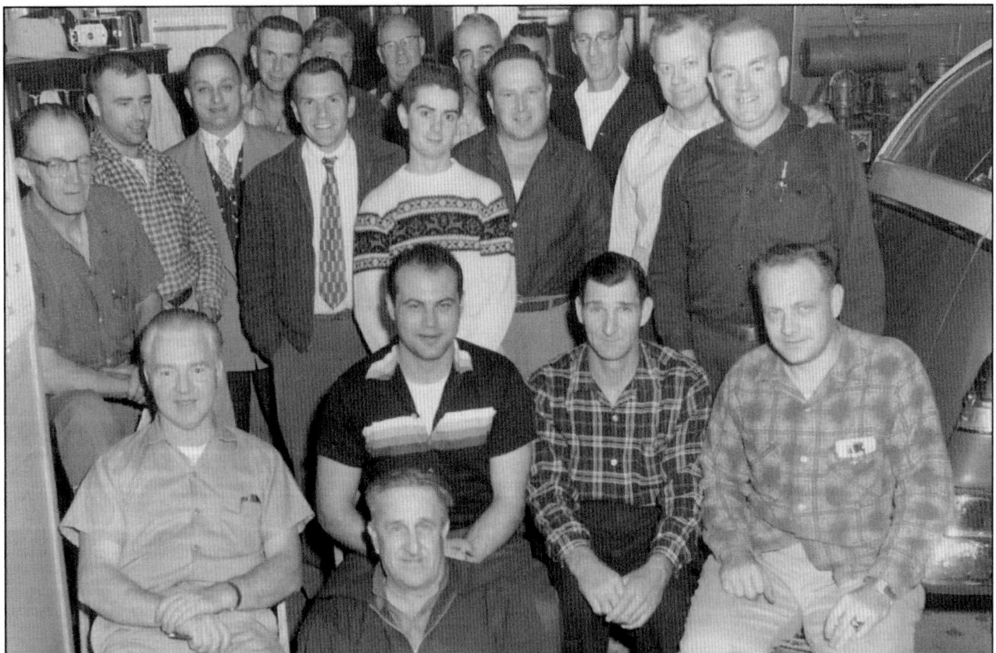

A 1952 photograph of volunteer firefighters includes Louis Skelton, seated second from the right, and his brother David, standing on the far left in the second row. The other men are, from left to right, (first row) Gus Berthume; (second row) Richard Howard, Lester McInnis, Louis Skelton, and Donald Sleeper; (third row) David Skelton, Wally Fredericks, Frederick Osbourne, Ernest Prouty, Albert Fay, Lester Anderson, Will Strickland, and Edward Howard; (fourth row) Kenneth Brown, unidentified, Al Cronin, unidentified, and Wilbur McIntire; (fifth row) Charles Bunton. (Courtesy of Albert Fay.)

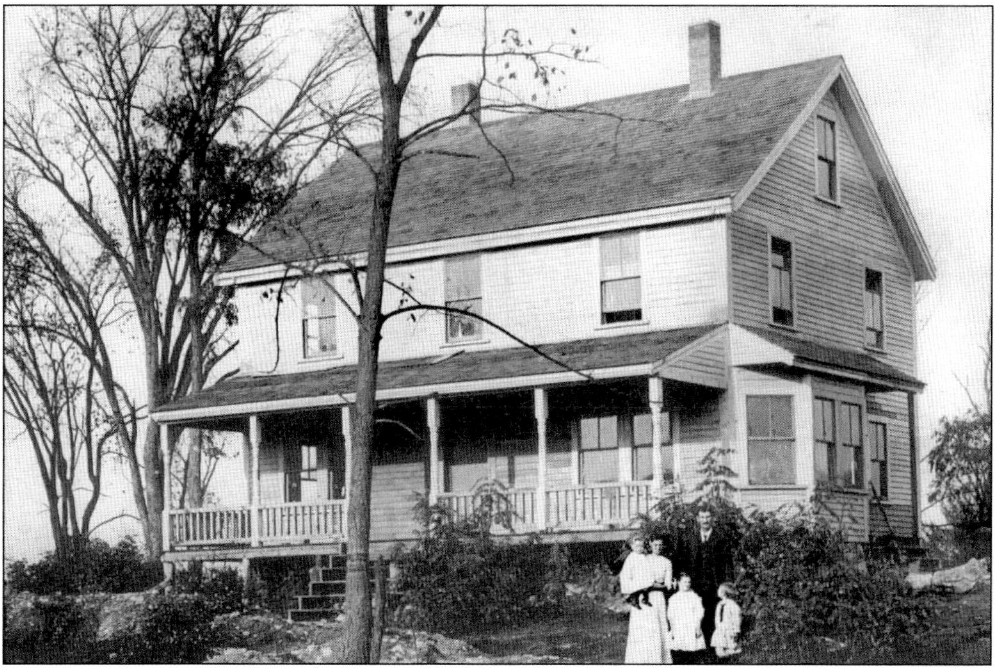

Horace Skelton and his young family are shown at his home on Francis Wyman Road. The house and farm have been occupied by members of the Skelton family since they were built in 1910. Iona Skelton Ganley, the daughter of Horace Skelton, shown in the arms of her mother, now occupies the house. (Courtesy of Burlington archives.)

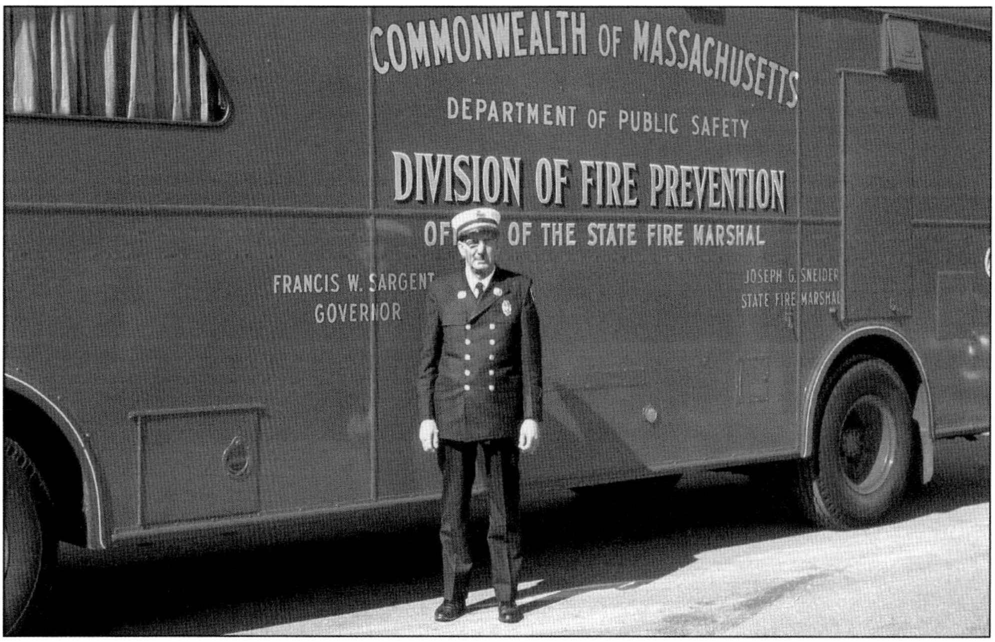

Louis Skelton became a permanent member of the fire department in 1955. He was named Burlington's Outstanding Citizen of the Year in 1977. The patriarch of the family, Louis Hamilton Skelton, was the last Skelton to serve on the fire department, retiring in 1978. (Courtesy of Burlington Fire Department.)

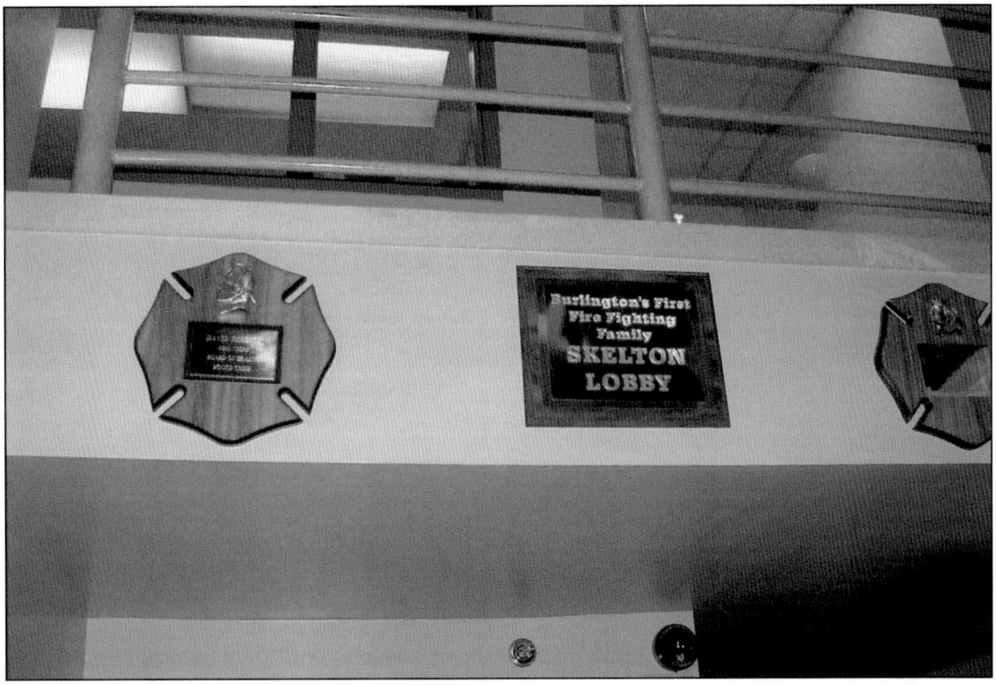

In 2005, the fire department dedicated the Skelton lobby at headquarters in honor of the six members of the Skelton family who served the fire department in the last century. They were respected, hardworking, and honest members of the community. (Courtesy of Toni Faria.)

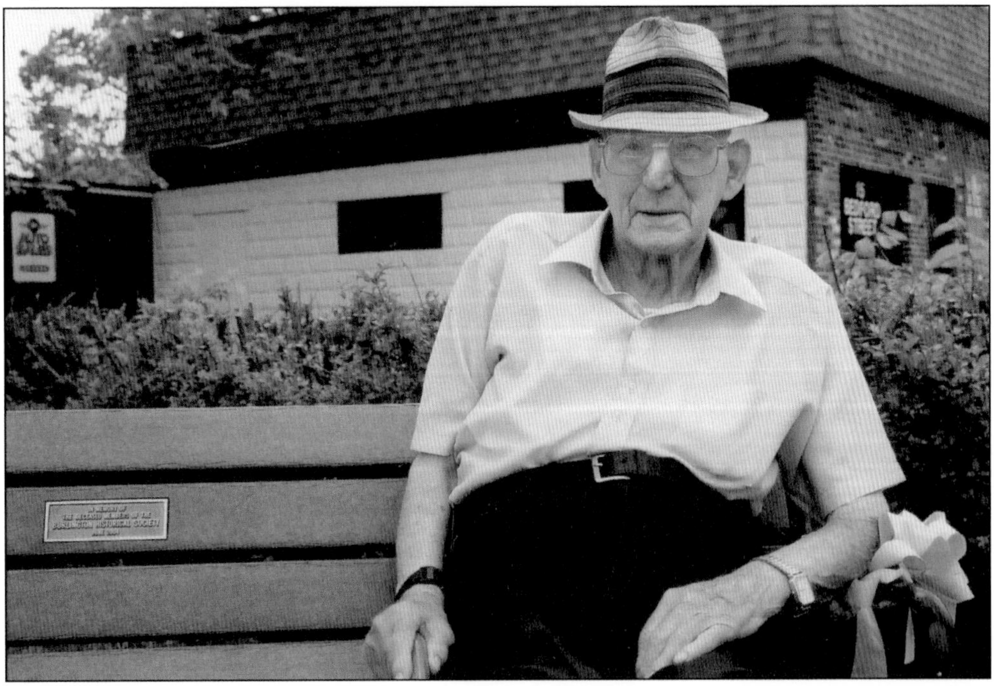

Louis Skelton is seated at the bench dedicated to historical society members at the historical museum. Louis was an active member of the Burlington Historical Society since its inception and was the treasurer for over 25 years. (Courtesy of Burlington Historical Commission.)

Eight

APPARATUS

The following 21 photographs of Burlington apparatus were provided by retired Lexington firefighter Robert Washburn. His collection of fire apparatus photographs is a result of his knowledge and interest of fire department vehicles. His years in the fire service prompted his interest in the equipment, and he began photographing the vehicles and cataloging them. He is widely known for his knowledge and expertise of fire apparatus. The rapid growth of Burlington brought about the need for improved and more efficient equipment. Aerial ladder trucks were added. Rescue vehicles and the latest equipment continued to be purchased. The changes in the equipment over the years are shown in the photographs as the department grew and became the modern department it is today.

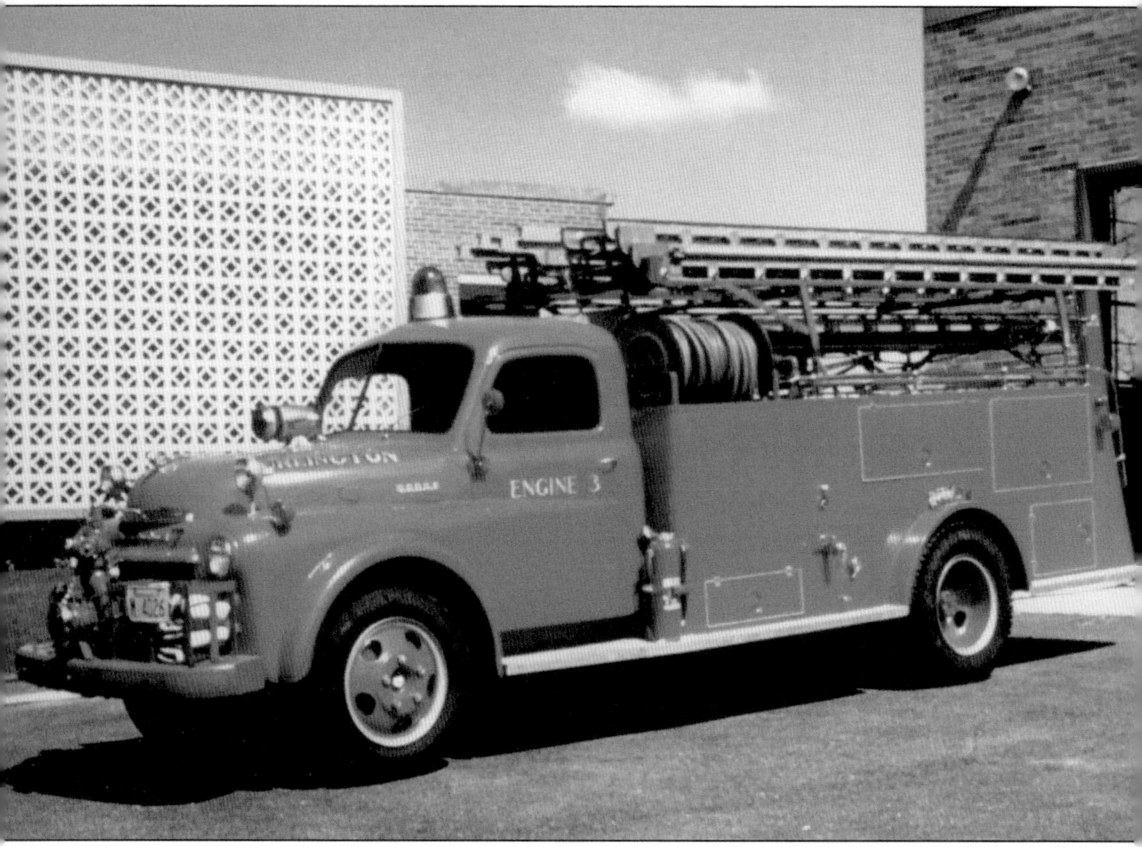

The 1949 engine 3 Dodge was built by the department; it was replaced in 1971 by a new American La France/Pioneer pumper and was given to the Massachusetts Fire Academy.

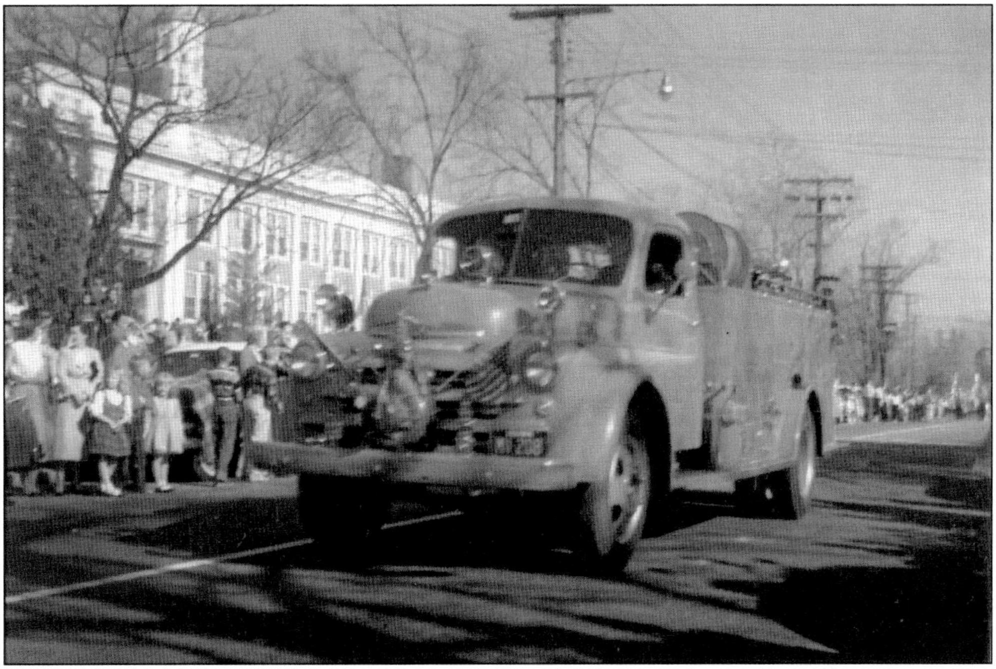

Seen is the 1949 engine 3 Dodge booster and hose truck. Volunteers built the 500-gallon front-mounted pumper with a wooden hose body. It was constructed of wood because of the metal shortage after World War II and the Korean War.

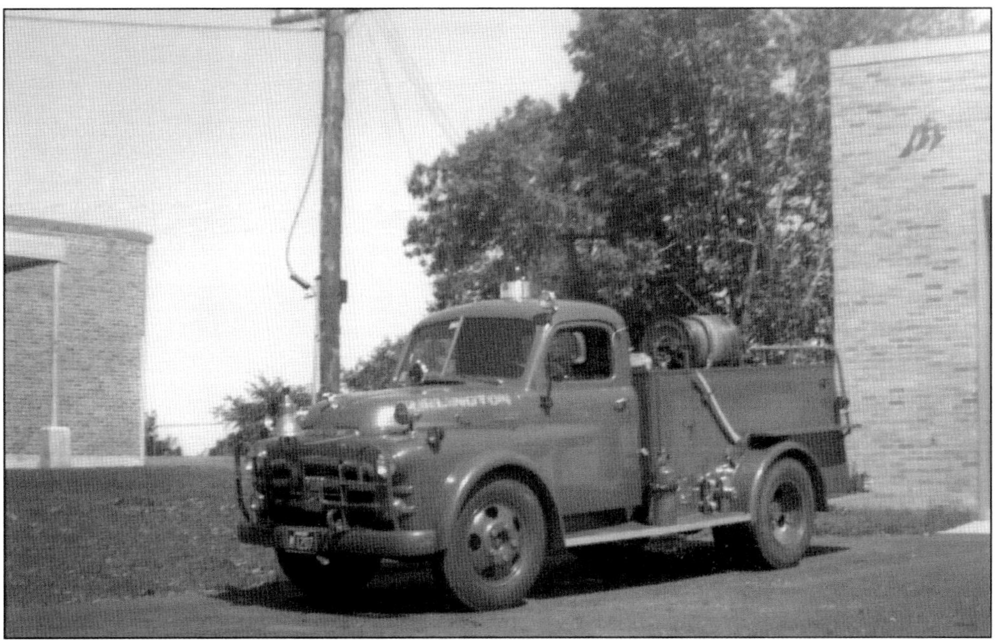

A 1949 Dodge that was built by the department appeared in a 1958 Lexington parade. It was made from a delivery truck donated by Reed Ham Works in Burlington. It was used for many years before being donated to the Massachusetts Fire Academy.

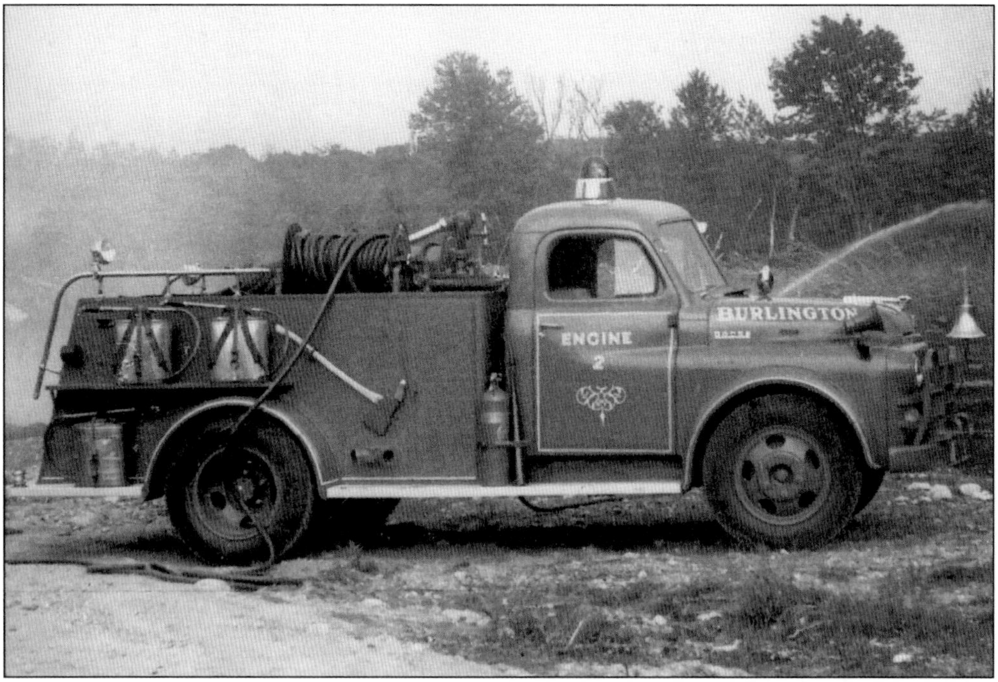

This 1951 Dodge booster and hose truck replaced engine 2, a 1930 Ford, which was a department-built truck that had been kept at Walter Skelton's farm from 1933 to 1937. It was the department's primary brush fire truck until replaced by a new engine 2, a 1969 Ford Farrar 500-gallon-per-minute front pumper.

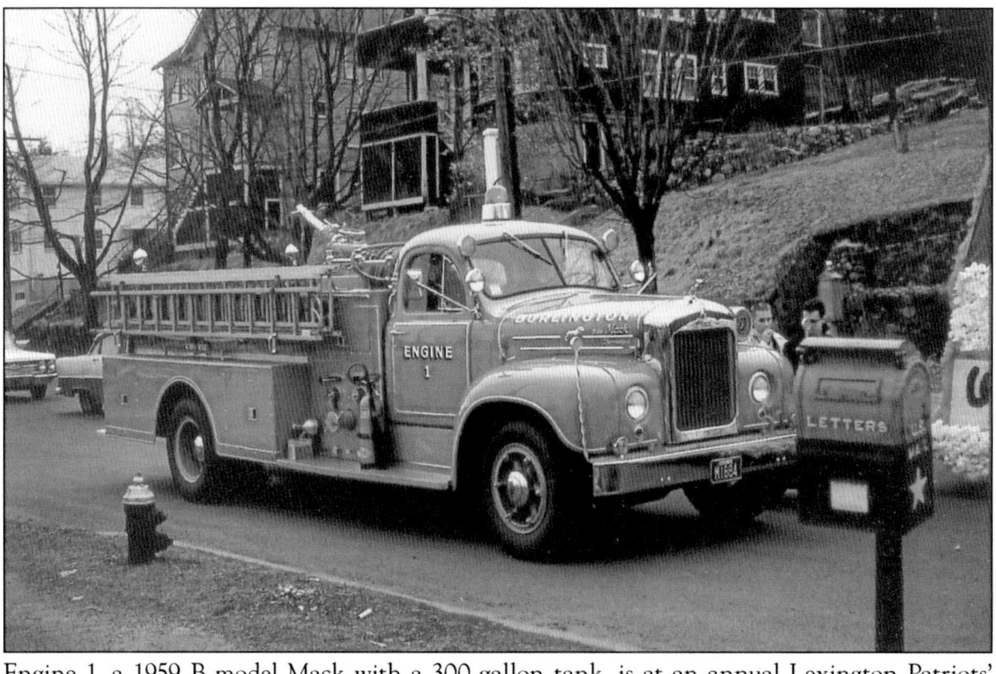

Engine 1, a 1959 B model Mack with a 300-gallon tank, is at an annual Lexington Patriots' Day parade.

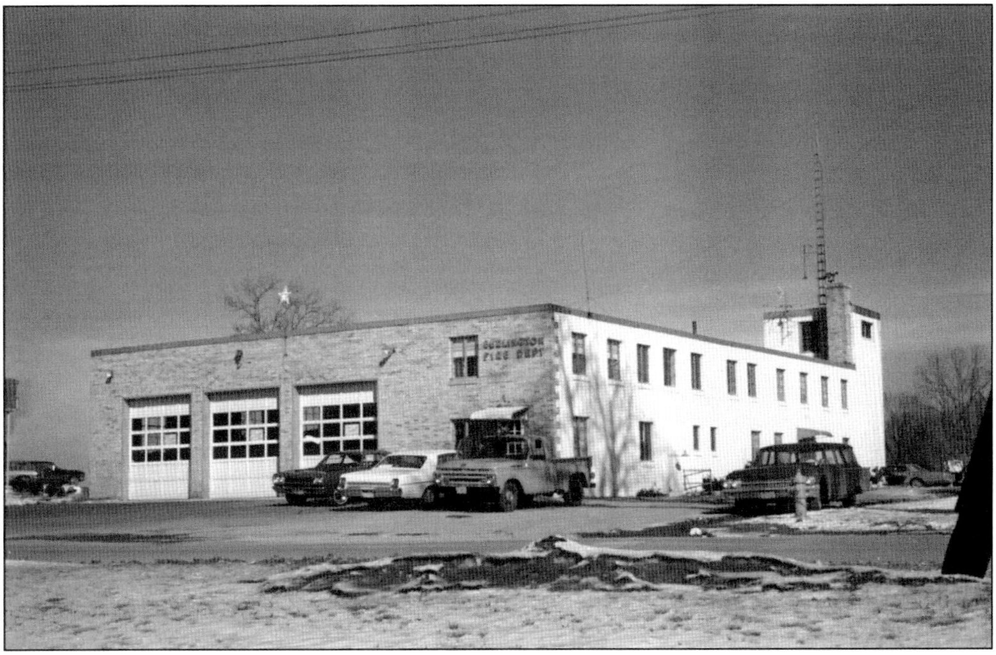

The newly built Burlington Fire Station in 1959 was the main headquarters for engines 1, 2, 3, 4, 5, and 6, the ambulance, a rescue truck, a fire alarm truck, the chief's car, a surplus power wagon, and a surplus jeep pickup.

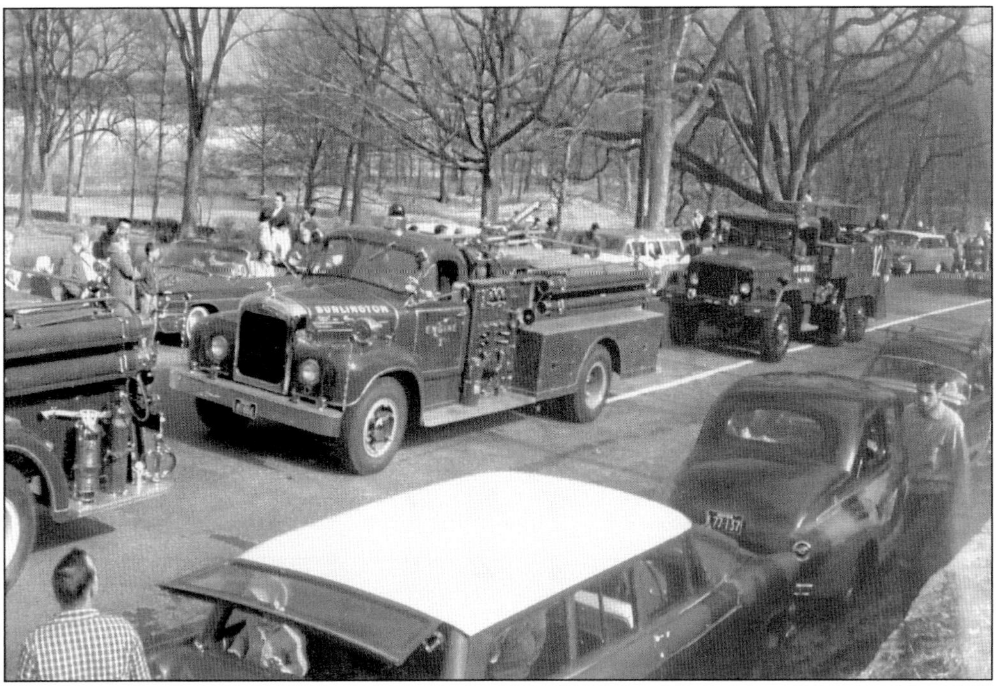

This is the 1959 Mack engine 1. It became the civil defense engine 10 from 1978 to the 1980s. The department retained this piece for a few years with hopes of refurbishing it as a parade piece. The funds were not available, and it was eventually auctioned off.

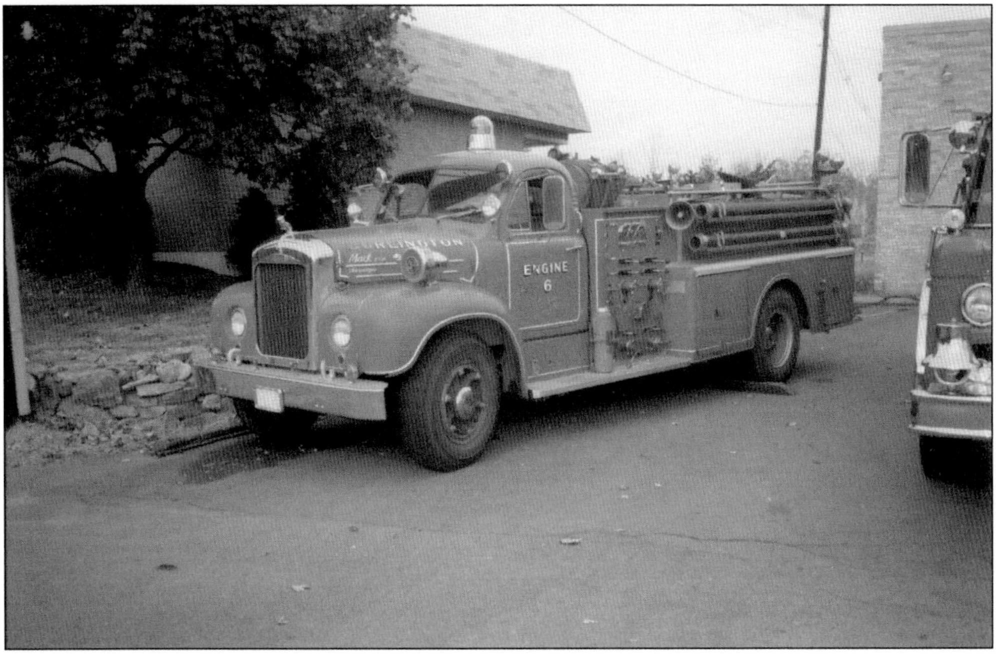

The 1962 engine 6 was a Mack 750. It was sold in 1990.

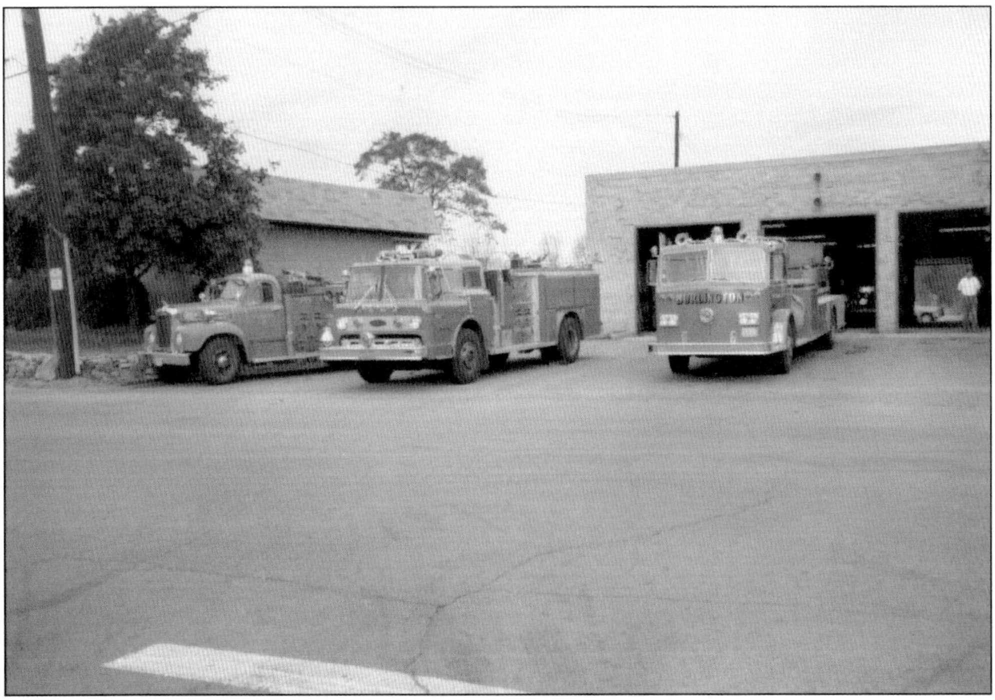
At main headquarters three vehicles are shown: engine 6, 1962; engine 1, 1983; and ladder 1, 1971, 85-foot Pirsch.

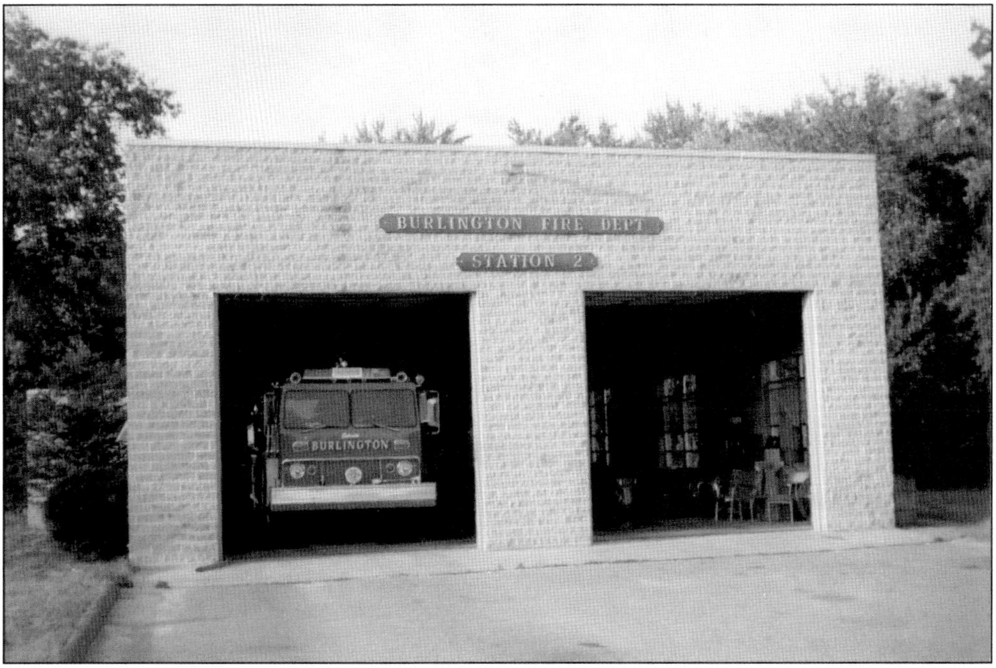

Built in 1971, the Burlington substation on Terrace Hall Avenue shows the 1978 Continental/Hendrickson, which later became the spare engine 3 in 1990.

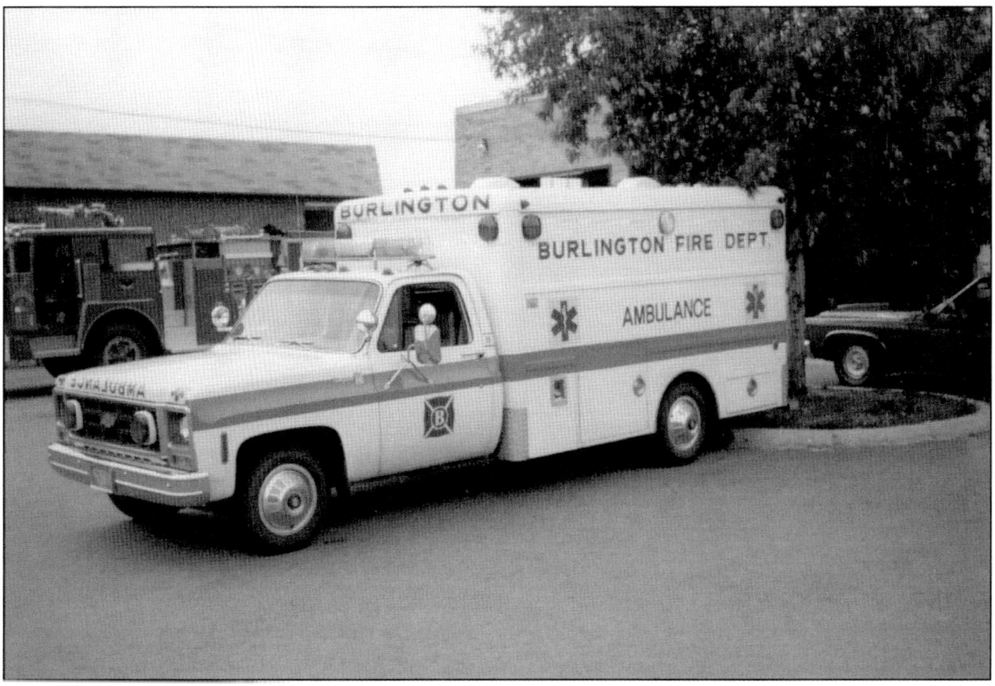

This 1981 fire department ambulance became a backup ambulance and was removed from service in 1987. It was later transferred to the Public Works Department for its use.

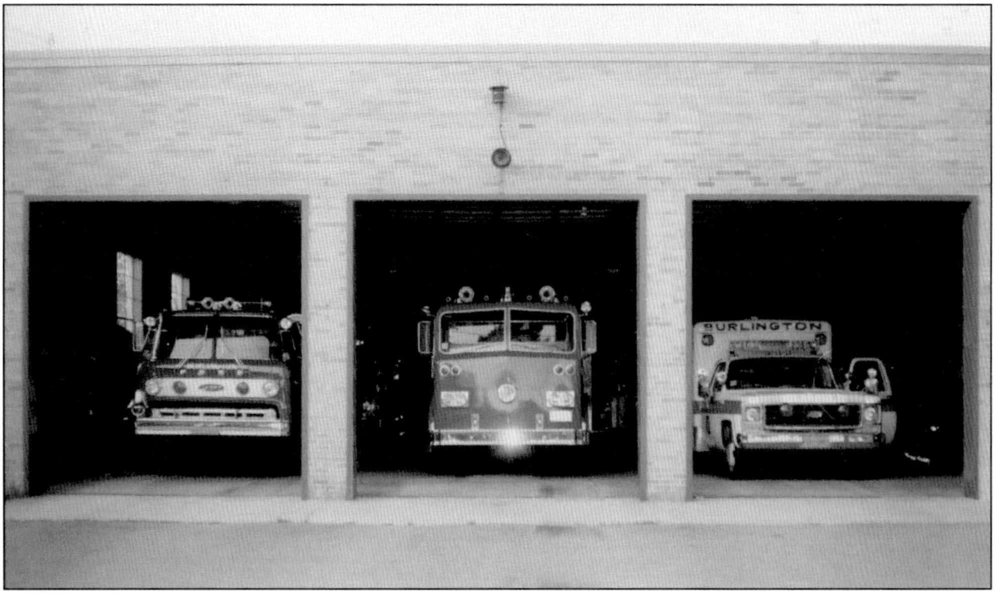

Shown at headquarters are engine 1, 1983; ladder 1, 1971 (a Pirsch 85-foot mid-mount aerial); and a 1978 Chevrolet ambulance.

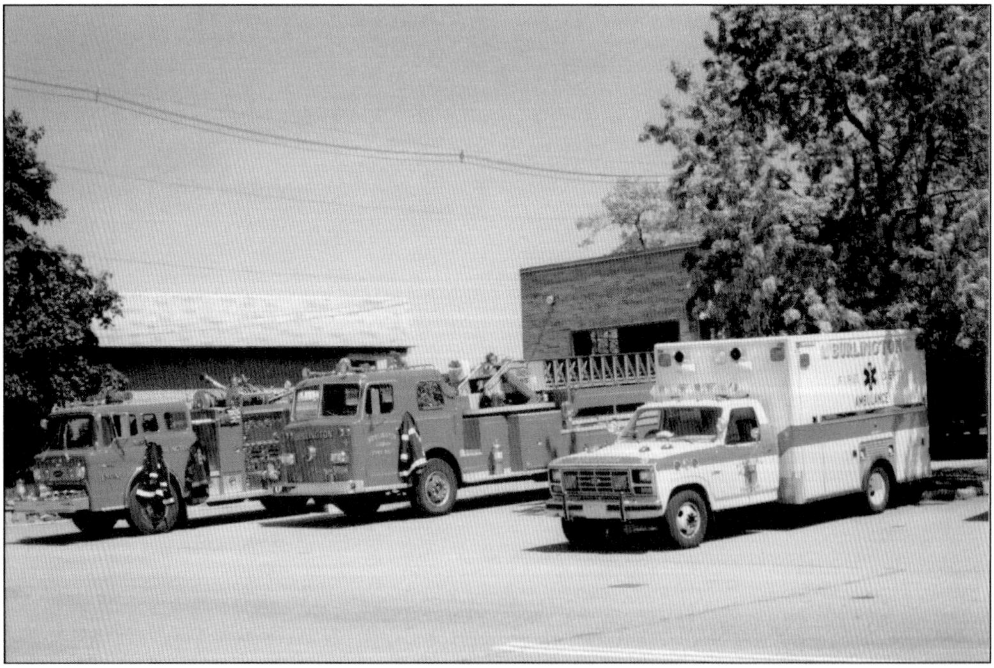

In 1985, at main headquarters, three pieces of equipment are shown. Engine 1, a 1978 1,000-gallons-per-minute pump, which went to the substation as engine 3; ladder 1, a 1969 Pirsch 85-foot mid-mount aerial, Burlington's first ladder truck; and rescue 1, a 1985 Ford.

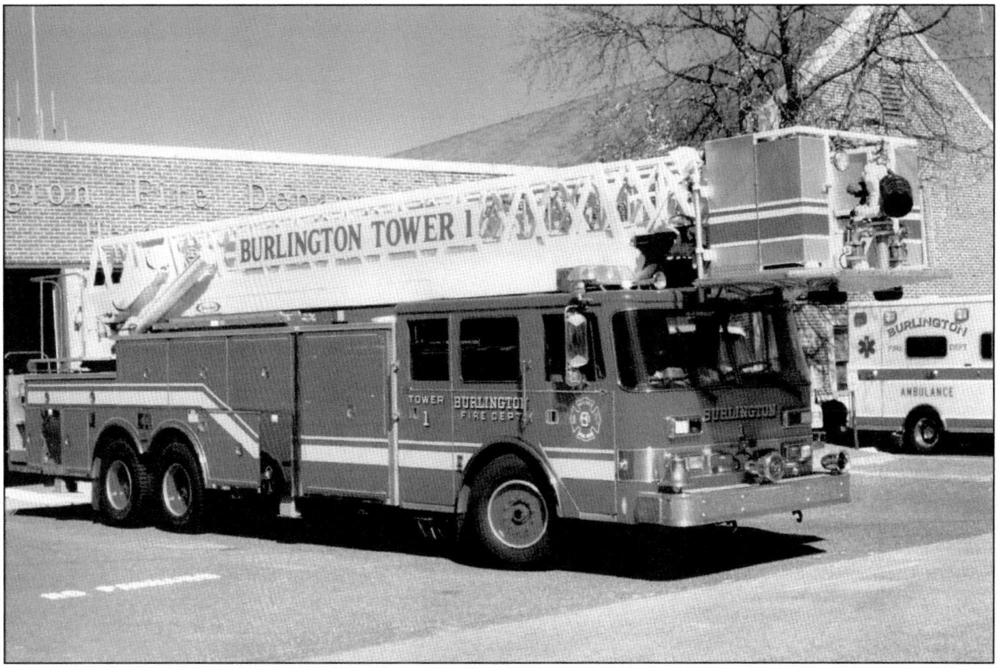

This 1990 Pierce aerial tower truck was Burlington's second ladder truck. Previously the department relied on ground ladders off either the pumpers or a ground ladder rack that was installed on engine 4 and on engine 3, a 1949 Dodge pumper.

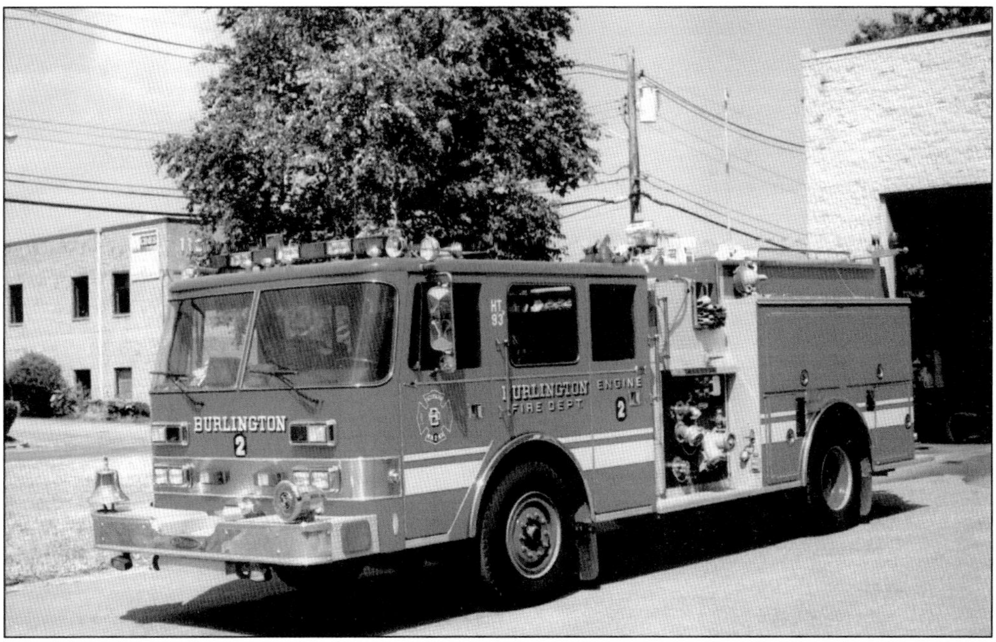

Engine 2 at the Terrace Hall Avenue substation was replaced there by engine 1, a 1998 Pierce/ Quantum. In 1996, the department completely rehabbed two of its pumpers. The former engine 4 was a 1975 International/Farrar 750 gallons-per-minute, 500-gallon pumper. Three different companies did the work.

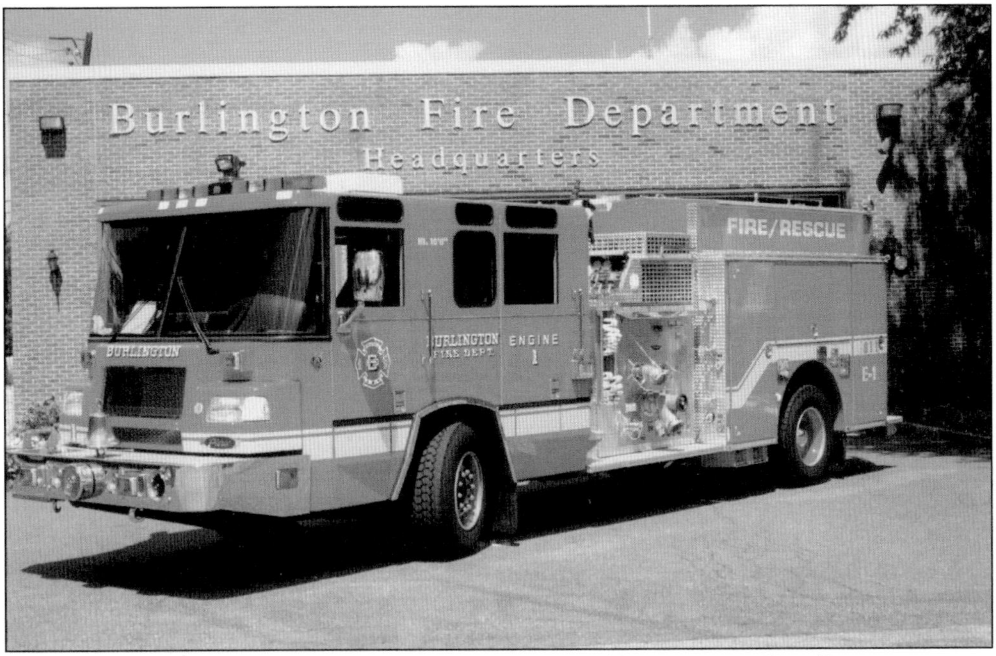

The new engine 1 at headquarters is a 1998 Pierce/Quantum 1,250-gallons-per-minute, 500-gallon triple combination pumper. It replaced the 1990 engine 1, which went to the Terrace Hall Avenue substation.

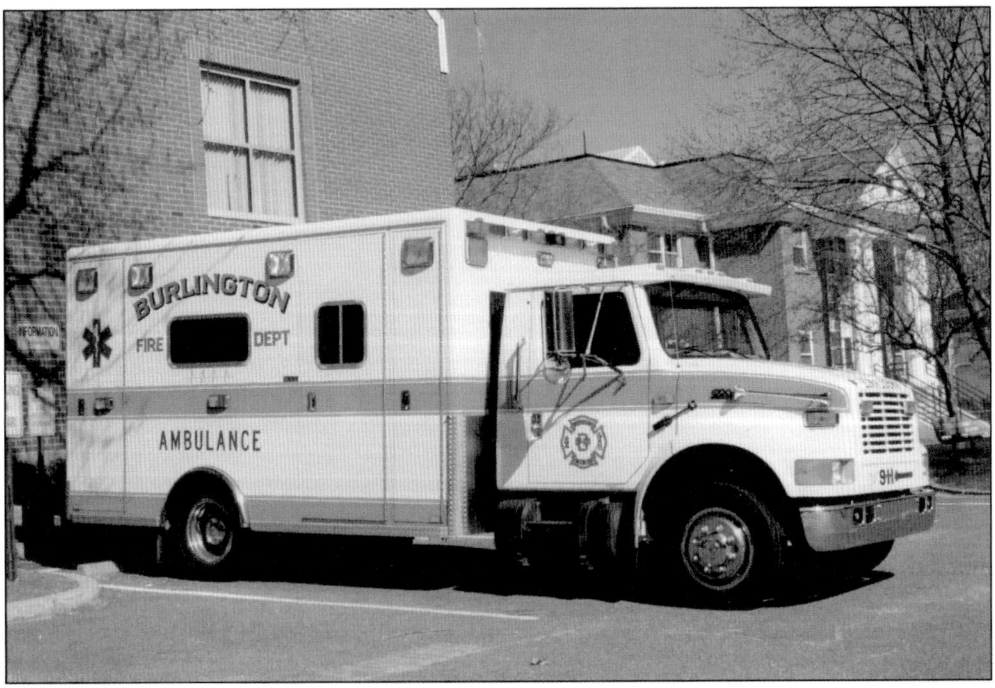

A new rescue 1 arrived in 2000. It was an International/Horton modular-style ambulance. It later became rescue 2 and was cross manned by a two-man crew from engine 3, or when manning is down, off the aerial tower.

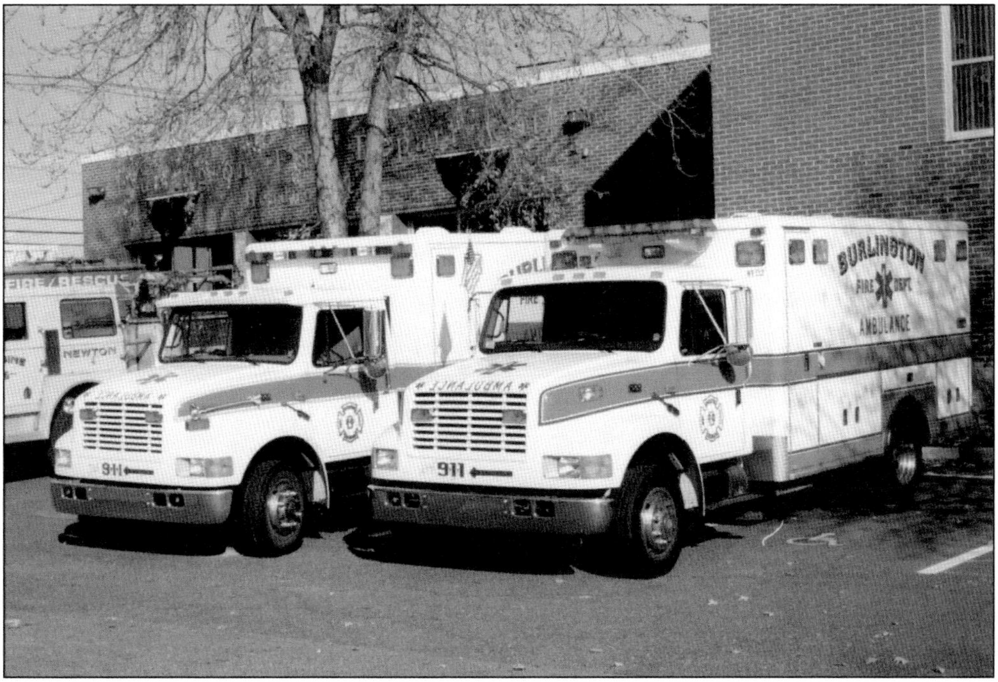

Photographed in 2001 are rescue 1 and 2. In 1953, the Burlington Fire Department took over the ambulance service using the police ambulance, a 1952 Cadillac. It had replaced the 1940s first police department ambulance. By 1989, the department began running two ambulances.

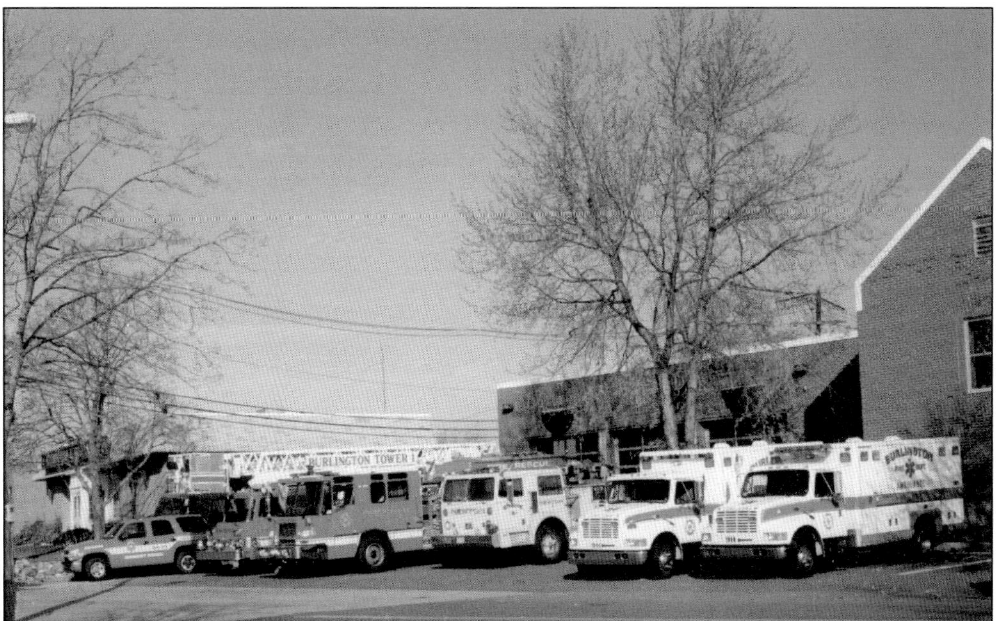

Pictured from left to right are a 2001 Chevrolet shift commander car; a 1990 Pierce aerial tower truck; a 1998 Pierce Quantum; a borrowed Newton, Massachusetts, vehicle; rescue 1, a 2000 International/Horton; and rescue 2, a 1994 International/Horton.

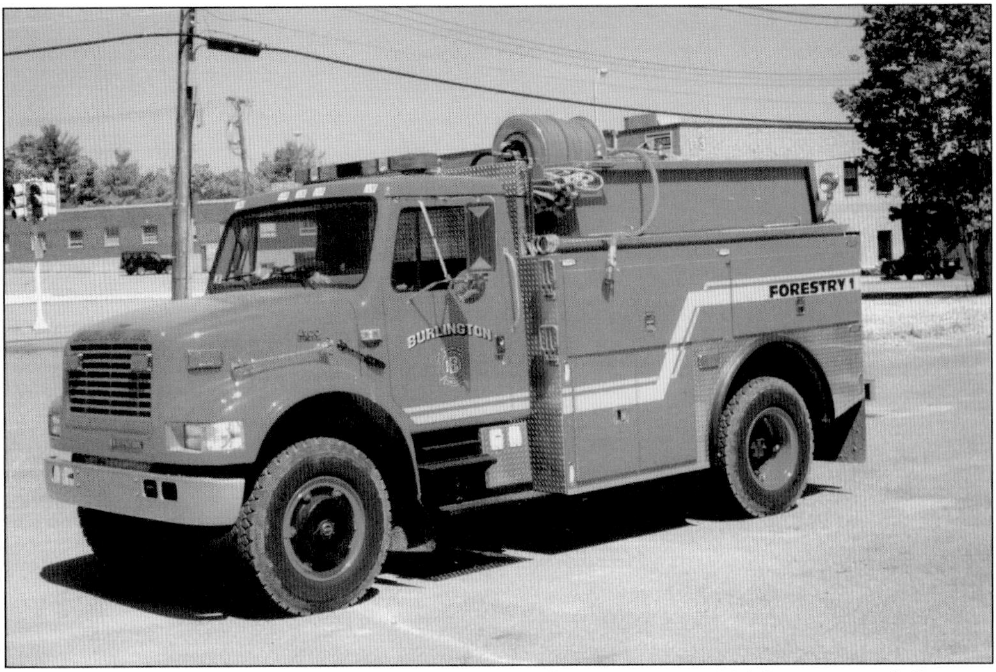

The forestry truck at the substation on Terrace Hall Avenue is a 1996 International Conway vehicle. It has a hard suction hose for drafting and is equipped to prevent overheating.

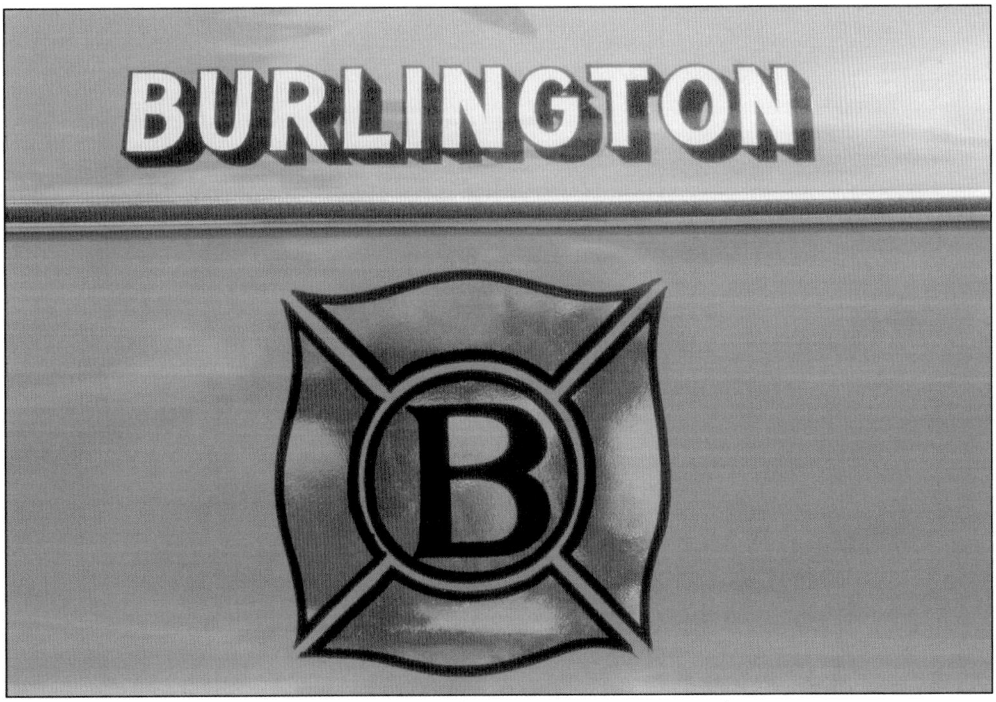

This is a Burlington decal. (Courtesy of Burlington Fire Department.)

Nine

Extra Duties, Activities, and Community Involvement

The members of the Burlington Fire Department not only serve the community as firefighters but also are actively involved in community activities. They help with fund-raising efforts for muscular dystrophy, cerebral palsy, the American Cancer Society, and other local charities. They are aware of the needs of the community and give their time and effort when needed.

The Burlington Fire Fighters Union Local 2313 has been generous in sponsoring baseball, softball, and hockey teams, cheerleader groups, and scholarships for graduating students. They have managed to volunteer for activities at the high school and host functions for senior citizens at the senior center.

The color guard has become a group to be proud of and an asset to the department. It is very well received in the community. It is partially funded by the fire department and the Burlington Fire Fighters Union Local 2313. The men of the Burlington Fire Department have given freely of their time to interact with the community. Their interest in community activities has been appreciated by all who have benefited from their generosity. The lobby of fire headquarters is filled with letters of appreciation and plaques from the many recipients of the department's generosity. The Burlington Quilters Guild donated a handcrafted hanging quilt in appreciation of the department's response to the 9/11 tragedy. The following photographs illustrate what the men have contributed over the years.

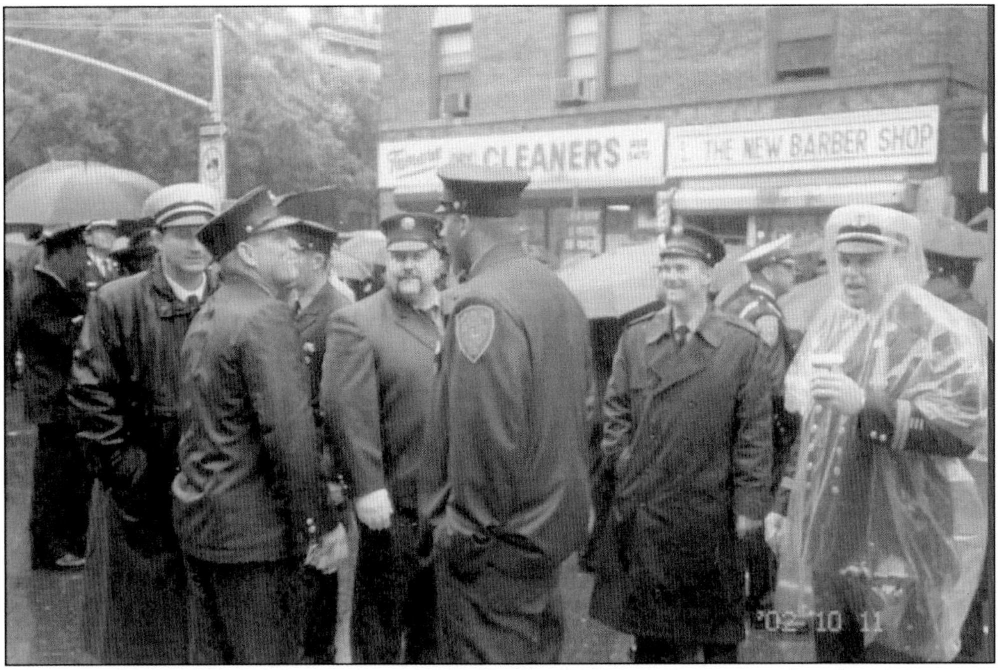

Burlington Fire Department personnel worked at the World Trade Center after the 9/11 tragedy in New York. Burlington firefighters joined firefighters from around the country to attend the funerals. From left to right are Steven McLean, James Sherman, Clifford Comeau, Frederick Williams, John Price, and Kevin Browne. In addition, the firefighters also responded to the Gulf Coast after Hurricane Katrina in 2005. (Courtesy of Burlington Fire Department.)

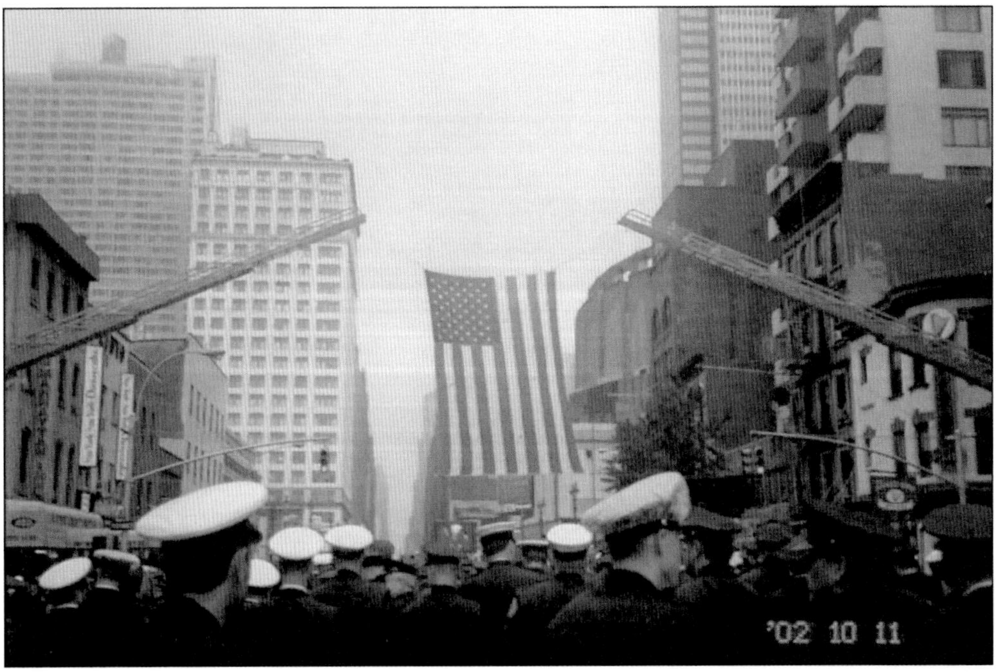

Burlington firefighters gather under the American flag at the World Trade Center after 9/11. (Courtesy of Burlington Fire Department.)

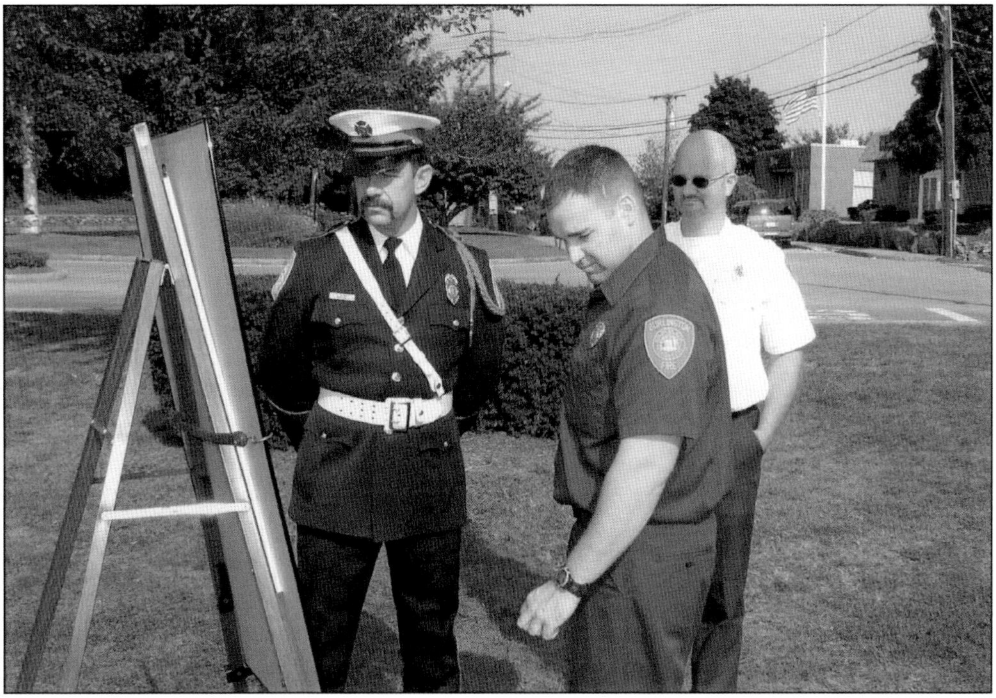

Assistant Chief Steven Yetman, James Hapenney, and Scott Carpenter are pictured here, from left to right, at headquarters preparing for the 9/11 ceremonies at the Burlington common. Assistant Chief Steven Yetman has risen in the ranks by his dedication and commitment to the fire department. He is a credit to the department. (Courtesy of Burlington Fire Department.)

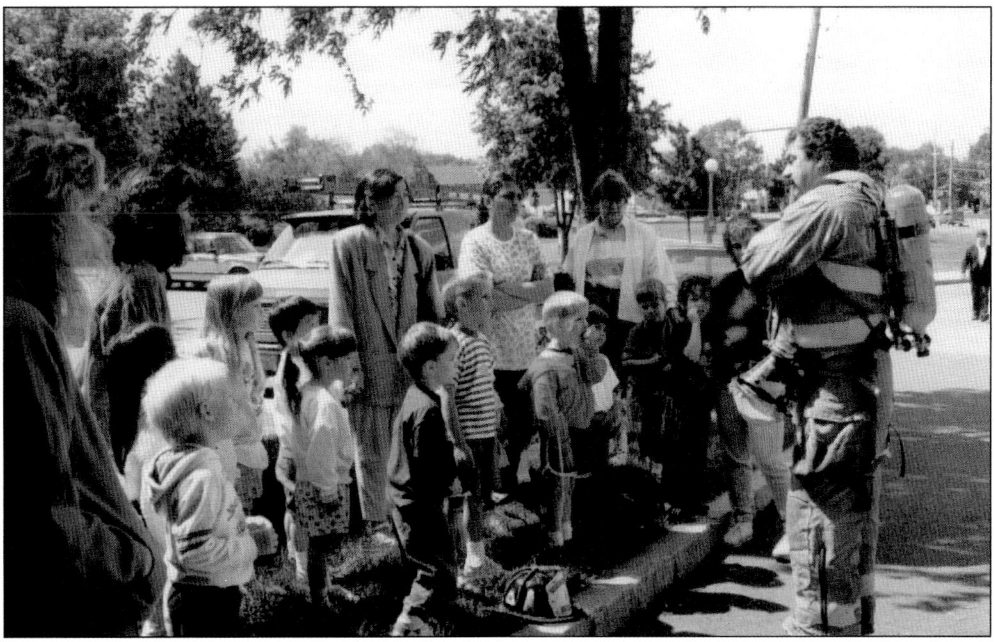

Firefighter Albert Brabant gives a demonstration to children on the clothing and duties of firefighters. Children are frequent visitors at the fire station, learning about fire safety and how the fire department works. (Courtesy of Paul Thibault.)

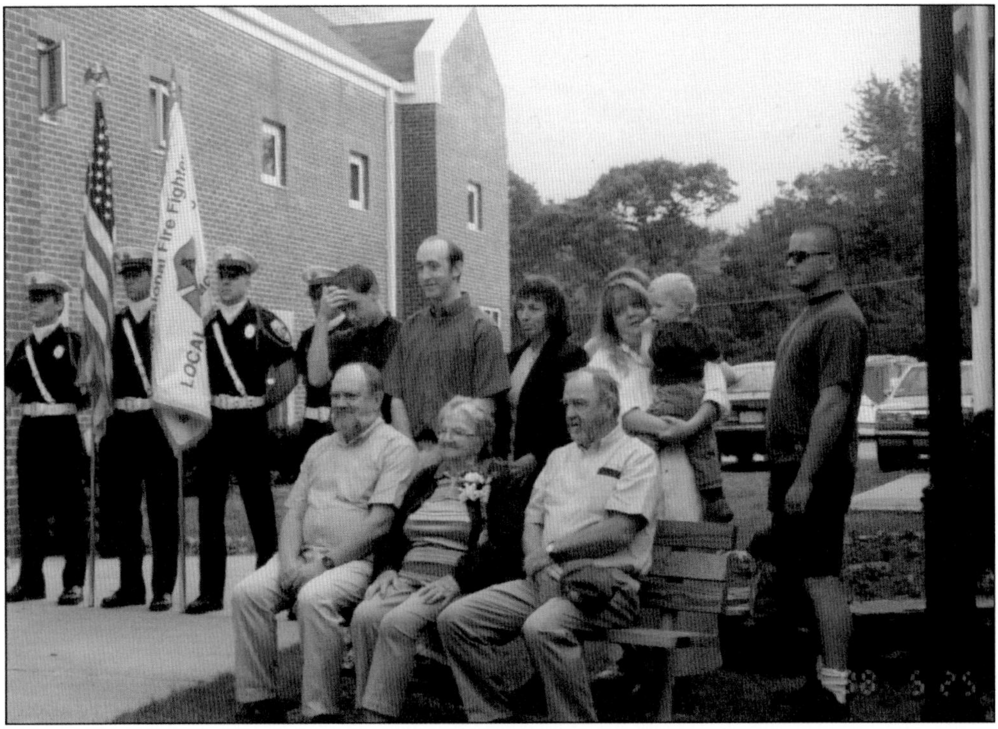

In 1998, a bench at the fire headquarters was dedicated to retired firefighter Edward Howard. He was a call firefighter in 1938 and appointed a full-time firefighter in 1955. He retired in 1976. His wife, sons, and grandchildren were present for the presentation at the fire headquarters. (Courtesy of Paul Thibault.)

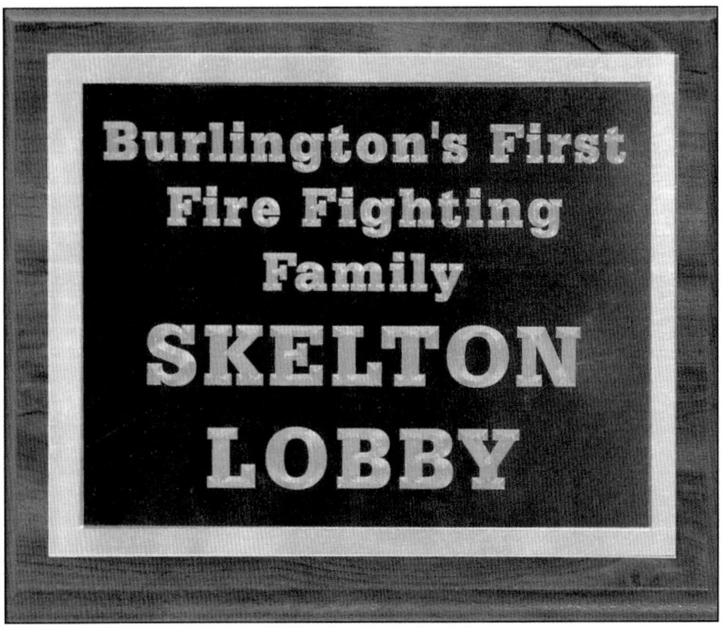

The Skelton lobby in the fire headquarters was dedicated in 2005 in honor of six members of the Skelton family who served the Burlington Fire Department in the last century. The dedication of the lobby of the fire headquarters in the name of Louis Skelton was held with members of his family and firefighters present. He was honored for his devoted service on the Burlington Fire Department. (Courtesy of Burlington Fire Department.)

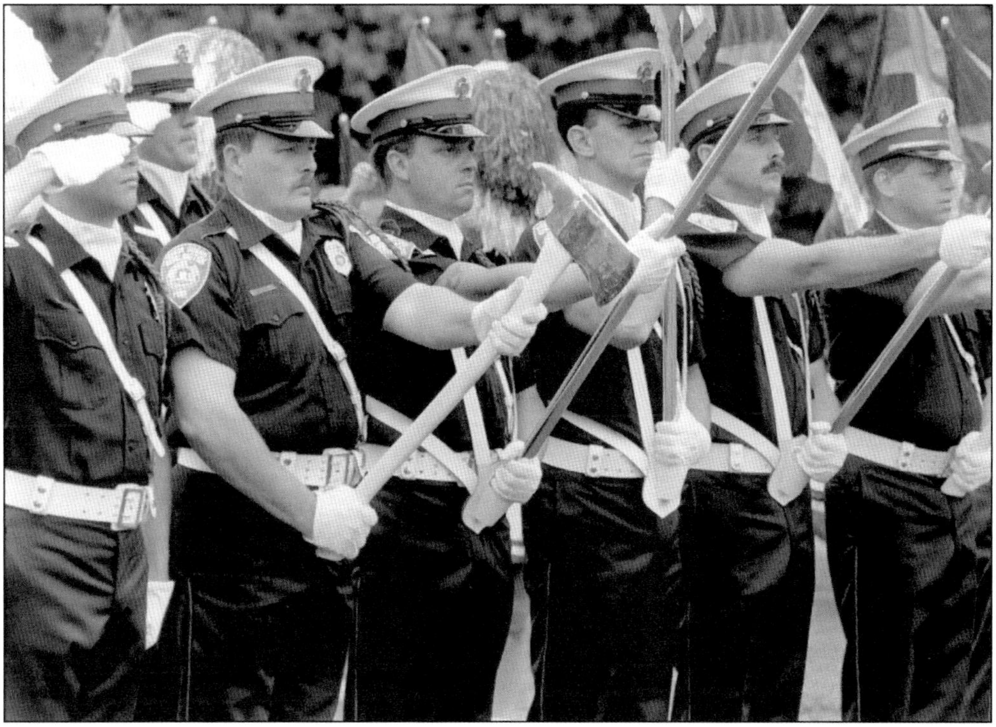

The color guard was formed in 1996 by firefighter Gerard Hanafin. The men who have volunteered to become part of the guard were trained by a member of the air force from the air force base in Bedford. The color guard provides a dignified and memorable memorial service. (Courtesy of Albert Fay.)

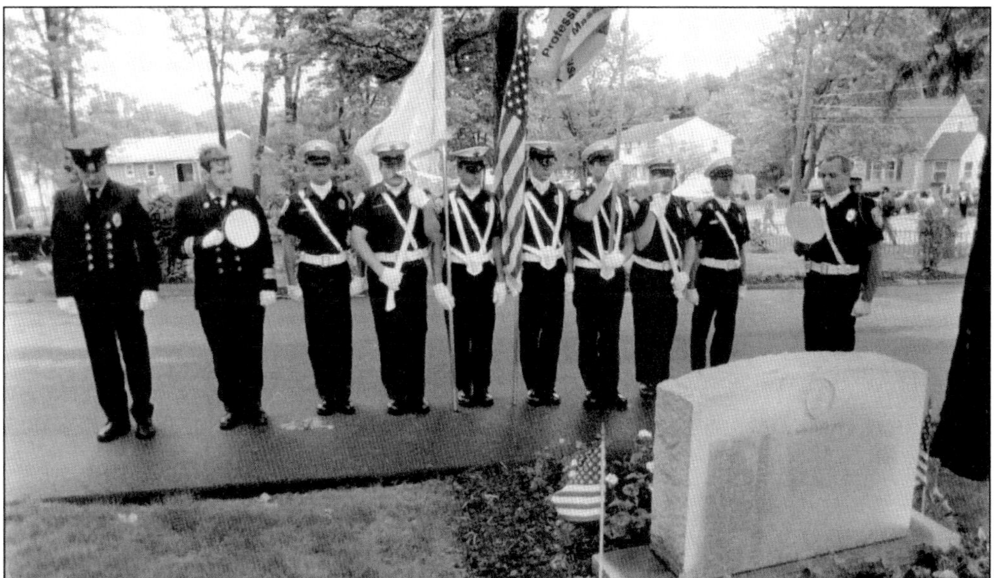

Color guard, once trained, began appearing at fire department funerals, veterans' functions, ceremonies, and dedications. They are well received and are commended for their excellent presentation. (Courtesy of Albert Fay.)

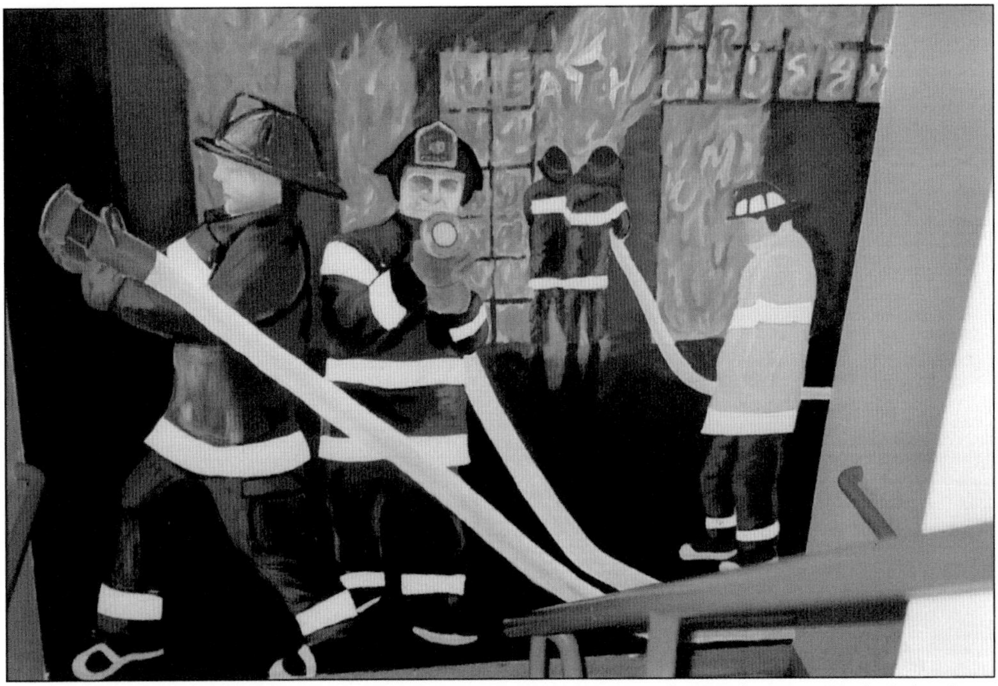

In 2005, the fire department was pleased to have artists from the Burlington High School art department come to fire department headquarters to paint a mural in the stairwell of the fire station. The murals are brightly painted action pictures of firefighters fighting a fire. (Courtesy of Toni Faria.)

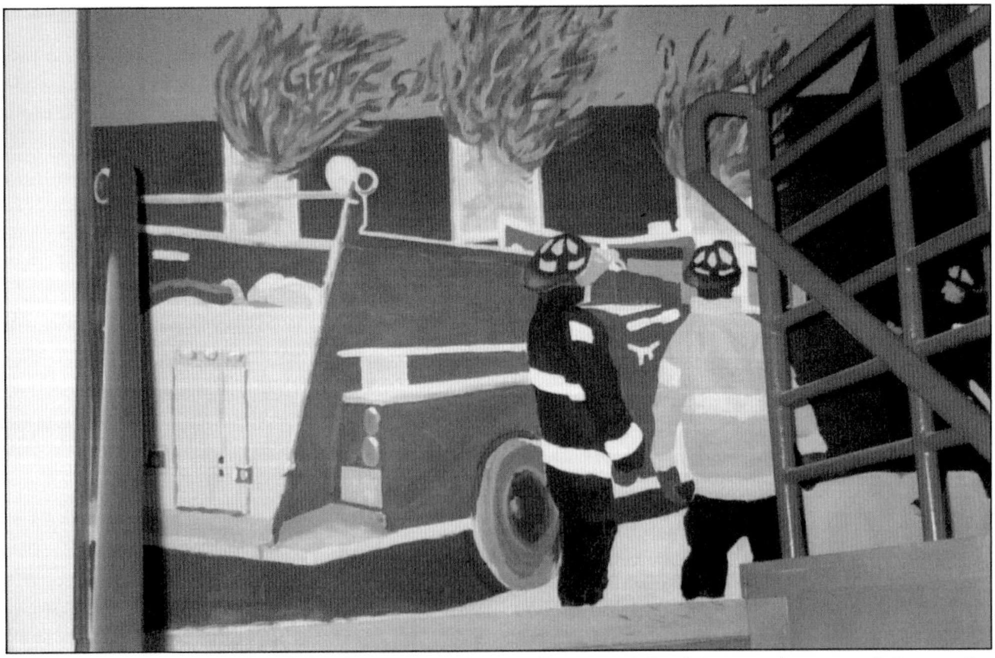

The murals on the walls of the fire station depict the actual firefighters who were fighting the fire. The artists were Michaela Baril, Tiffany Huang, Heather McCarthy, Chris McCabe, Kristen Rubino, and Geoff Silva. (Courtesy of Toni Faria.)

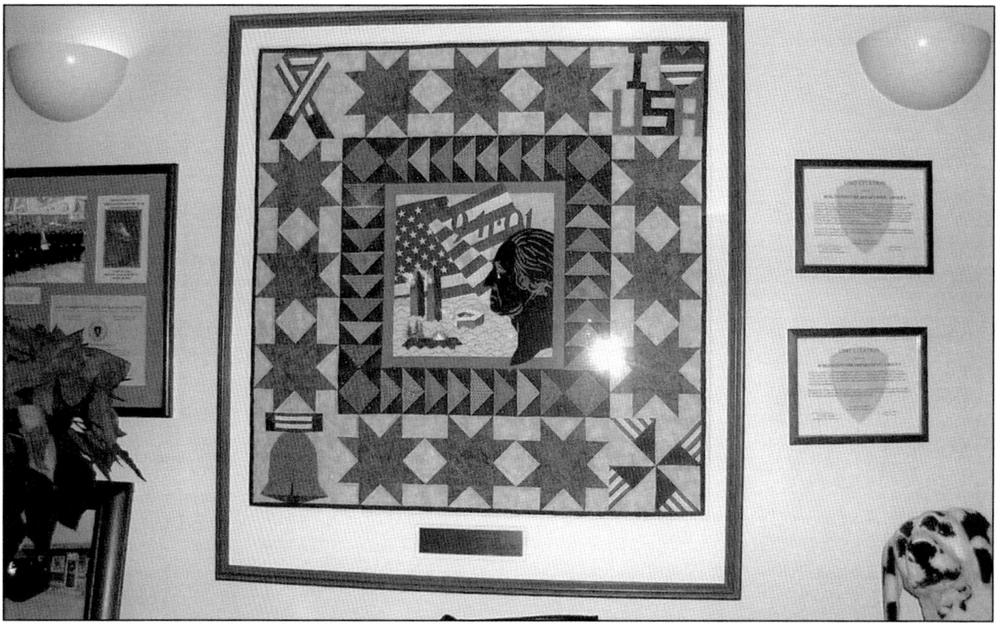

The Burlington Quilters Guild presented the fire department with a handmade framed quilt in appreciation of its involvement in the 9/11 emergency. It is on display in the lobby of headquarters. (Courtesy of Toni Faria.)

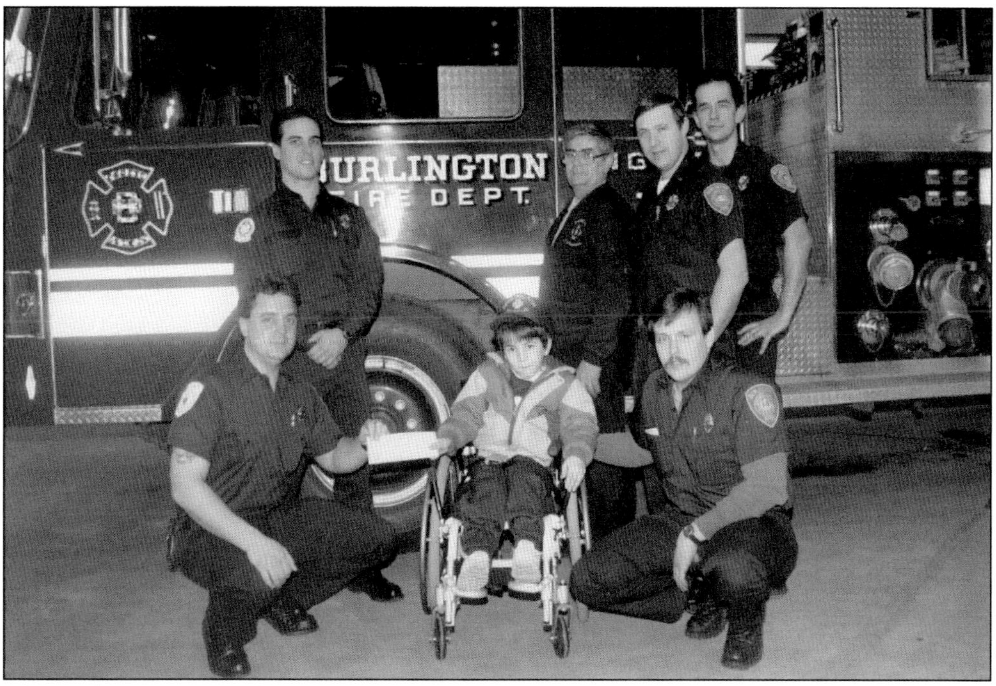

The Burlington Fire Fighters Union Local 2313 conducts fund-raisers, gives scholarships, and conducts the boot drive to benefit muscular dystrophy. In this 1991 photograph are, from left to right, (first row) Gerard Letendre, the muscular dystrophy poster child, and Edward McLean; (second row) Michael Bibbey, Albert Fay, Leonard Sawyer, and Steven Yetman. (Courtesy of Albert Fay.)

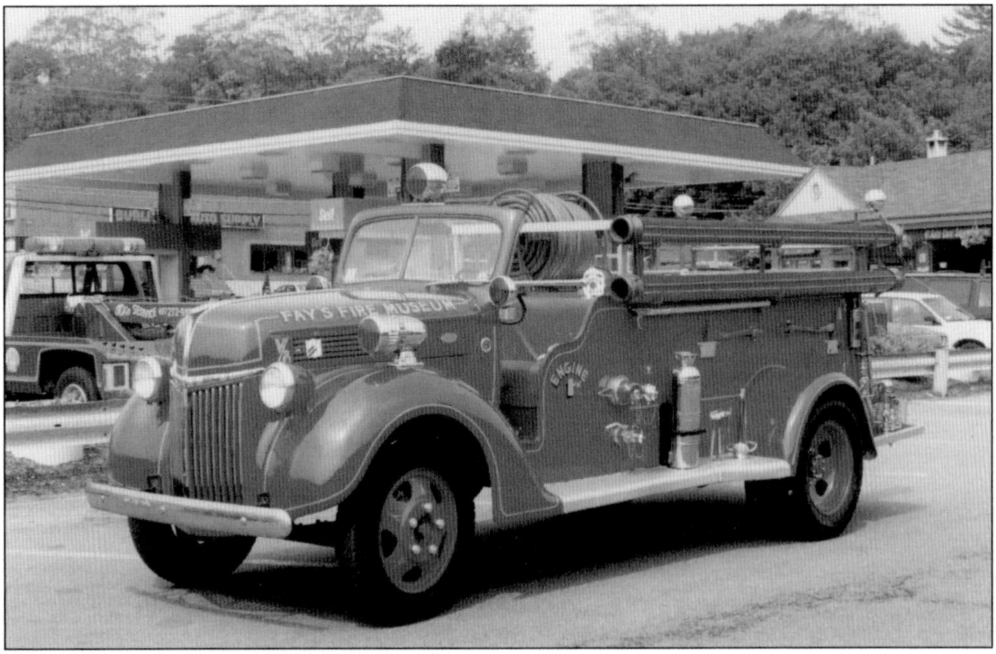

Retired firefighter Albert Fay continues his interest in the fire service by volunteering to display and drive his antique fire truck to promote local fund-raising and civic programs. He has been active in the community since he began his fire service career as a call firefighter. He was voted Burlington's Outstanding Citizen of the Year and is currently on the board of selectmen as well as serving on several other committees. (Courtesy of Albert Fay, photograph by Keith Butler.)

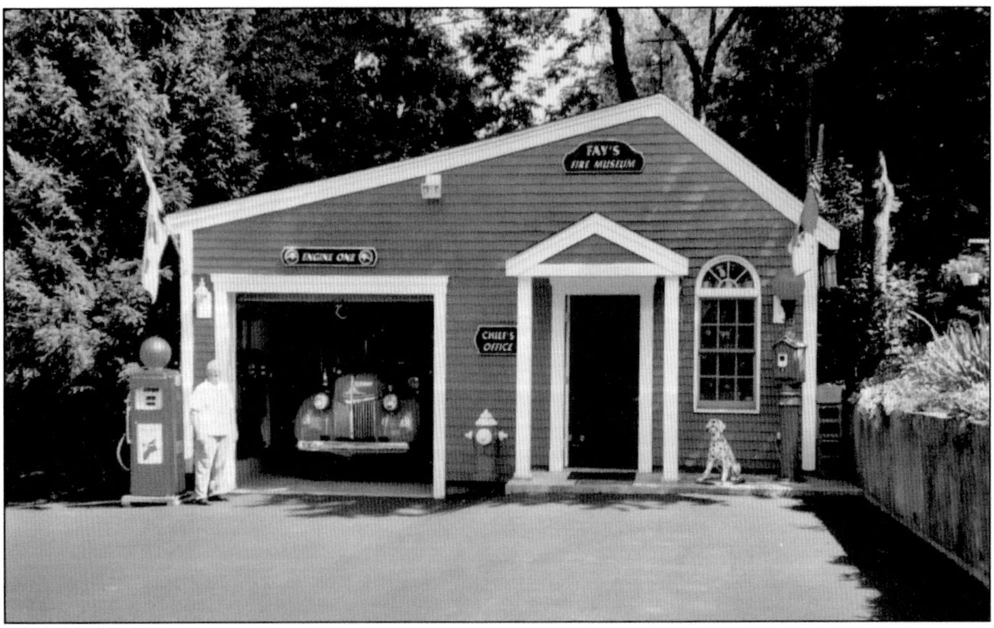

Albert Fay is the owner of Burlington's only fire museum. The museum is located on his property on Raymond Road. It is free, and tours are often conducted for interested parties. It contains a large collection of fire-related artifacts as well as an extensive collection of miniature fire vehicles. (Courtesy of Toni Faria.)

Three publications are connected to the Burlington Fire Department. *Protective Breathing Apparatus* by Lawrence Faria was published by Prentice-Hall in 1975. It is one of the first books written on self-contained breathing apparatus, used as a standard in fire academies across the country. The second is a children's book, *Fire Trucks*, written by local author Judith Jango Cohen, and features Burlington fire equipment and firefighters. *Town of Burlington, Mass. Fire Department* was written by Robert Washburn of Lexington. (Courtesy of Toni Faria.)

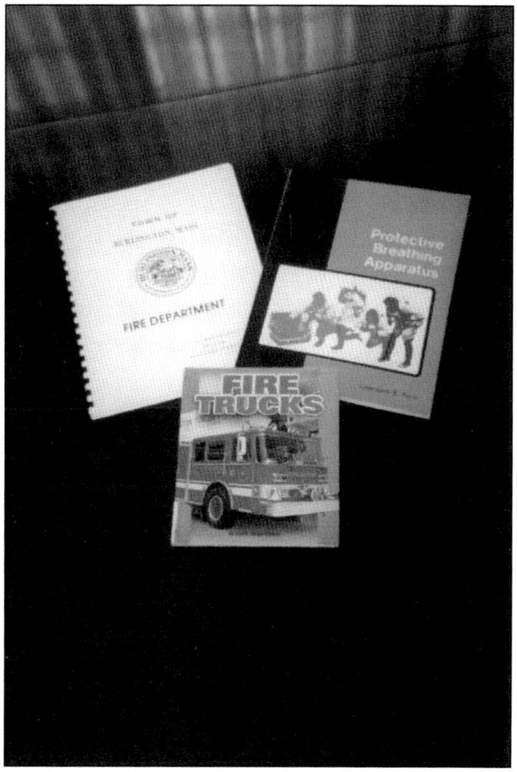

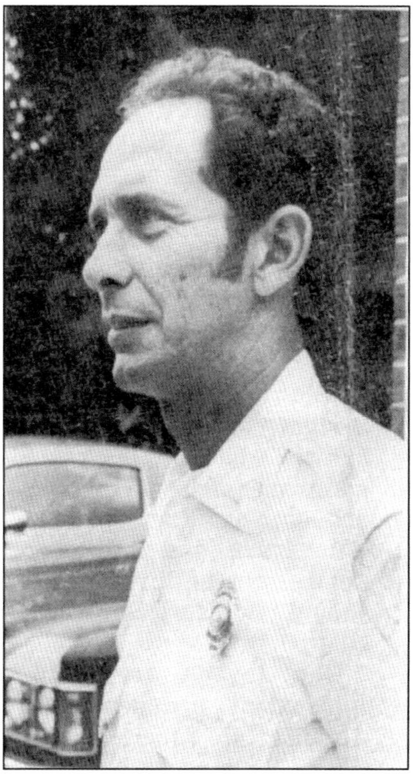

Lawrence Faria was an auxiliary firefighter for two years before being appointed a permanent firefighter in 1960. He was appointed a lieutenant in 1967, training officer in 1971, and fire prevention officer in 1981. He retired in 1984. He was an instructor for many years at the Massachusetts Fire Academy. (Courtesy of Toni Faria.)

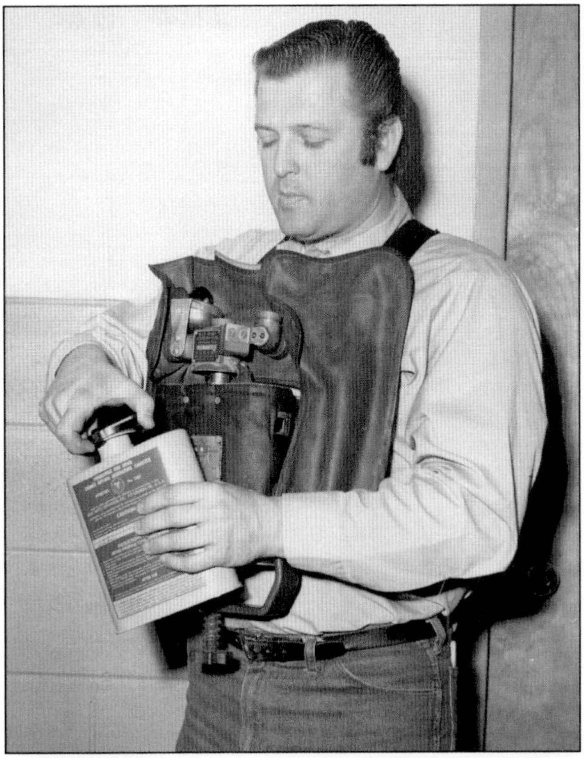

Burlington firefighter George Clark assisted Lawrence Faria in the demonstration of equipment for the *Protective Breathing Apparatus* book. (Courtesy of Toni Faria.)

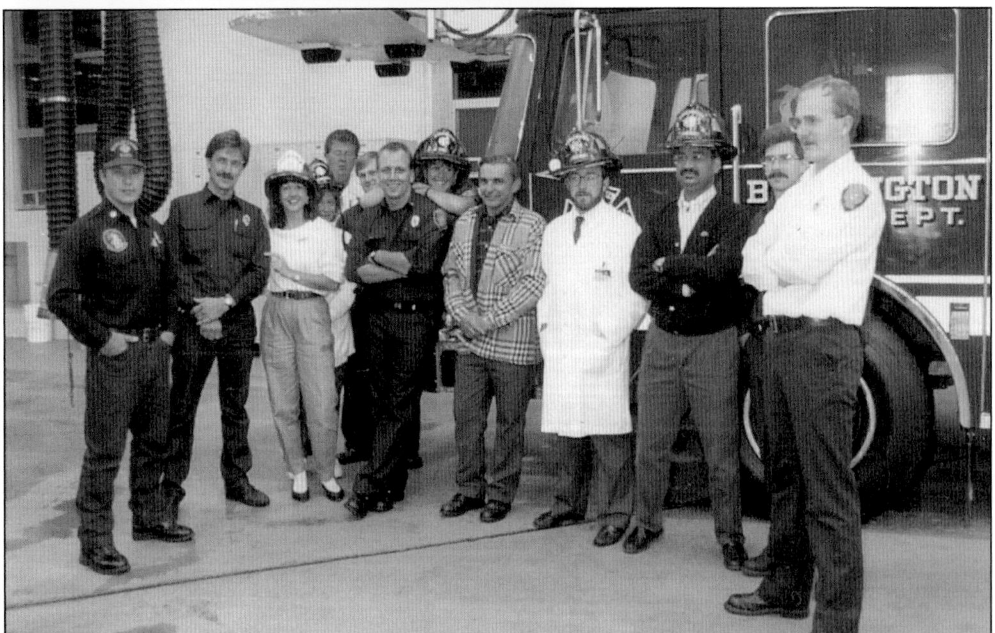

Firefighter William Bibbey, fourth from the left, is pictured in this gathering of volunteers. They coordinated a blood drive for Bibbey, who was seriously ill and in need of blood transfusions. It was one of the most successful blood drives ever held by the fire department. Scott Carpenter is on the far right. (Courtesy of Albert Fay.)

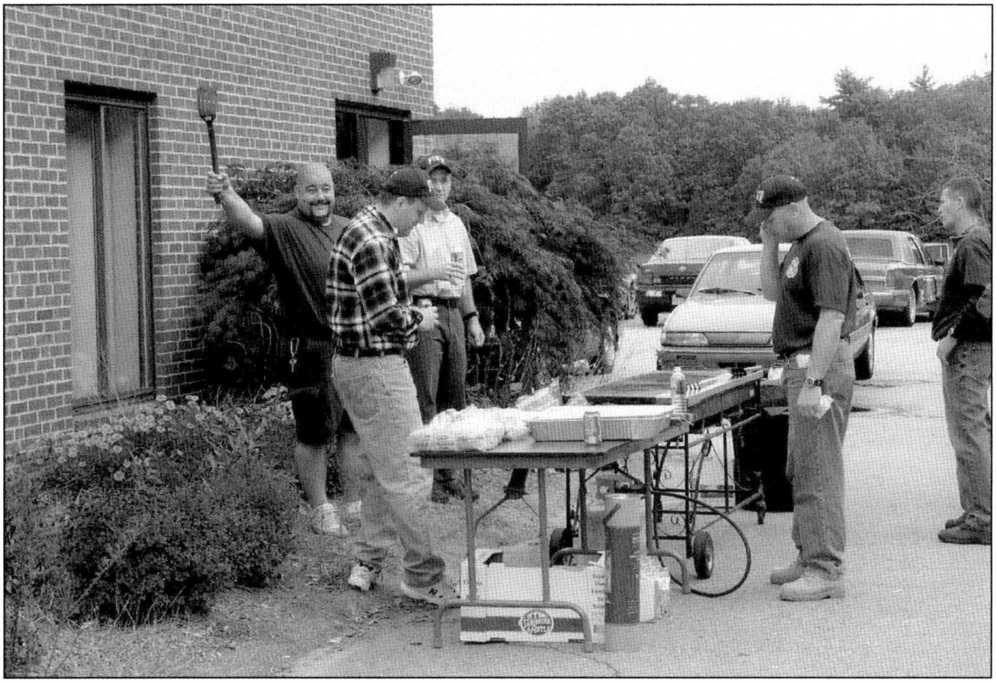

An outdoor luncheon at the senior center was hosted by the firefighters. Clifford Comeau (left) waves to the guests as he prepares to serve the seniors. (Courtesy of Burlington Fire Department.)

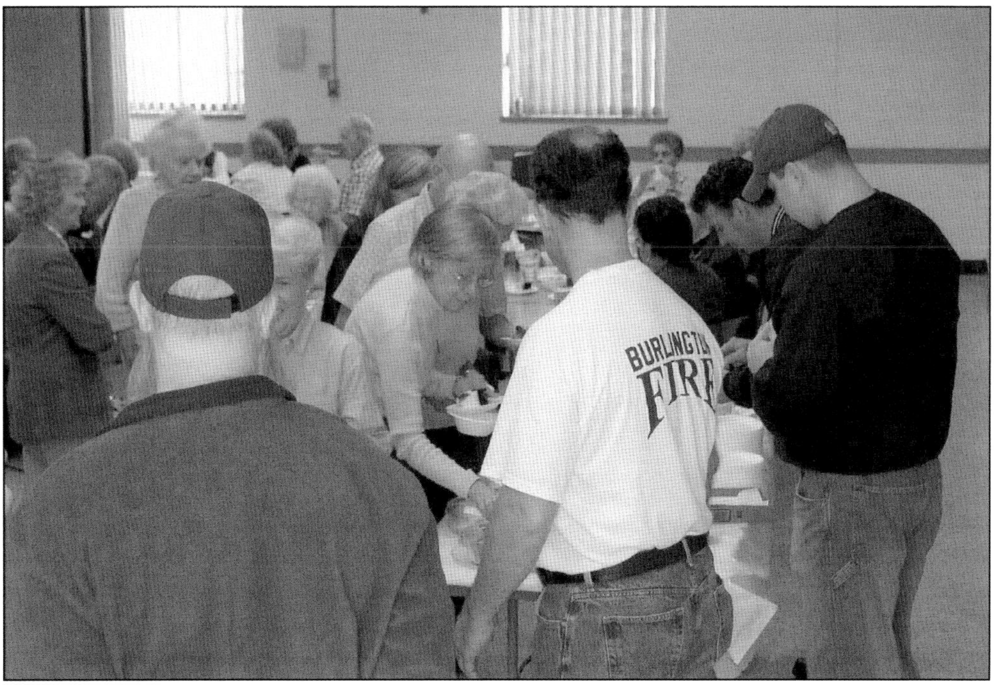

Burlington firefighters are shown hosting a dinner for senior citizens at the Council on Aging Senior Center. As part of their volunteer activities, the firefighters often host luncheons for the seniors. (Courtesy of Burlington Fire Department.)

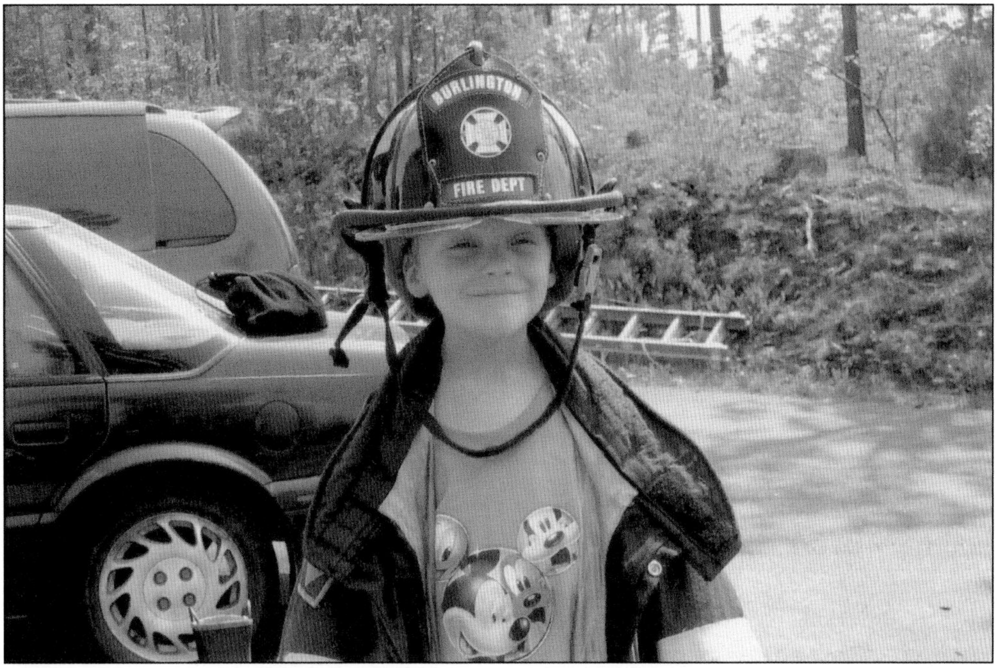

Michael McLaughlin, grandson of retired firefighter Paul McLaughlin and son of Michael McLaughlin, currently a firefighter, is posing in his fire hat. (Courtesy of Paul McLaughlin.)

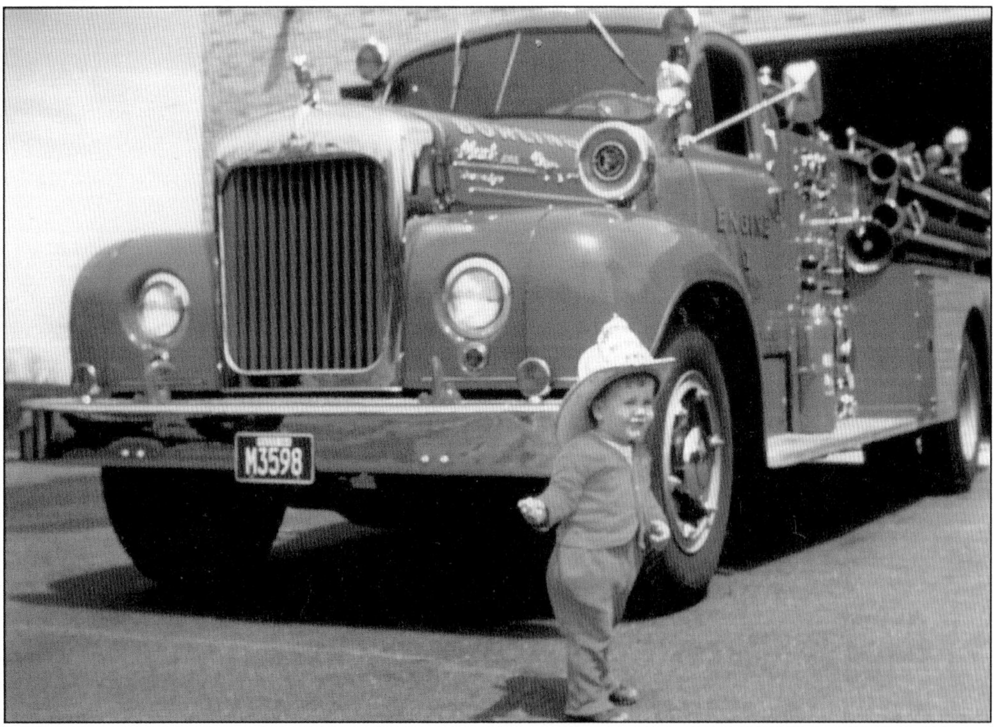

A toddler poses in front of his favorite place, the Burlington Fire Station. He grew up to be a firefighter. He is Capt. Scott Carpenter of the Burlington Fire Department. (Courtesy of Albert Fay.)

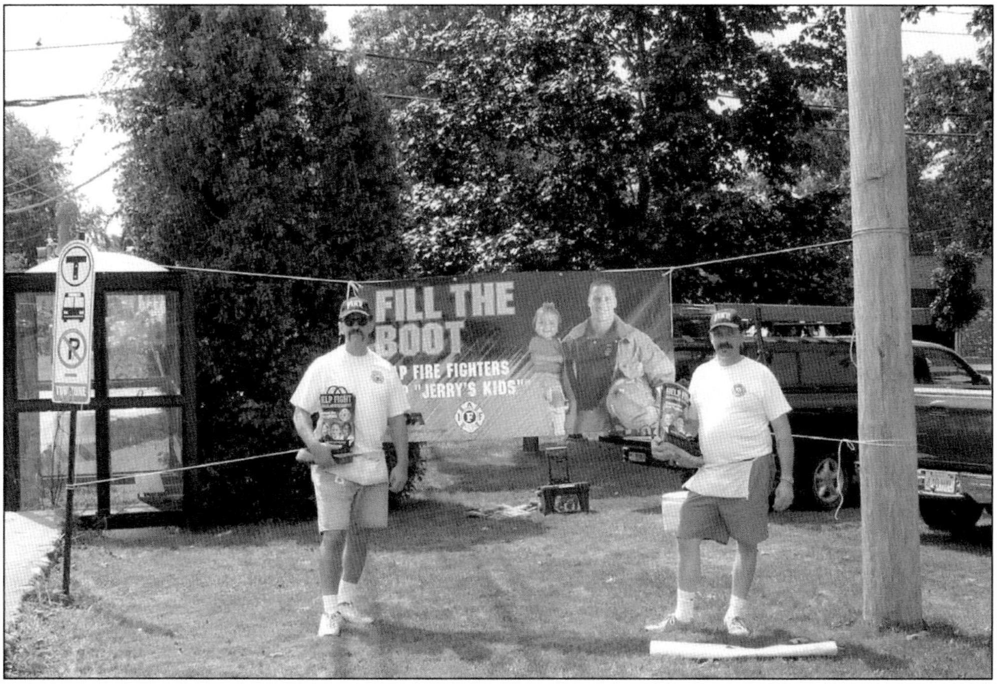

Firefighters Steven Yetman (left) and Paul Sullivan prepare for the boot drive for cerebral palsy, an annual event to raise money for the charity. (Courtesy of Burlington Fire Department.)

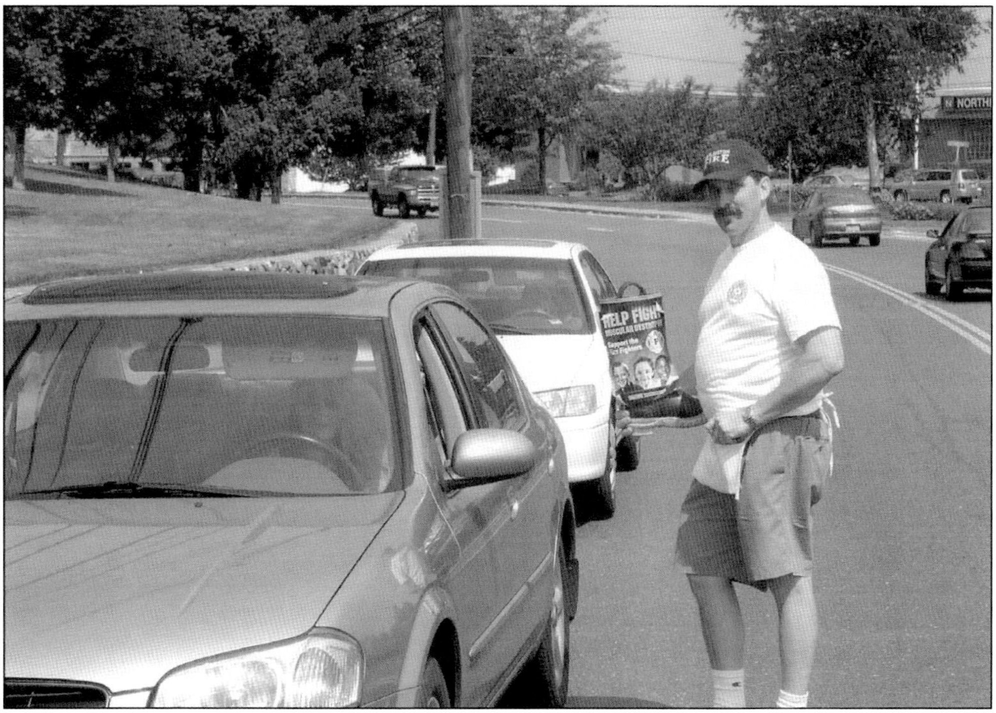

Paul Sullivan is shown here on Route 3A collecting donations from motorists for the annual drive for the cerebral palsy fund-raiser. Burlington holds an outstanding record for the money collected for the drive. (Courtesy of Burlington Fire Department.)

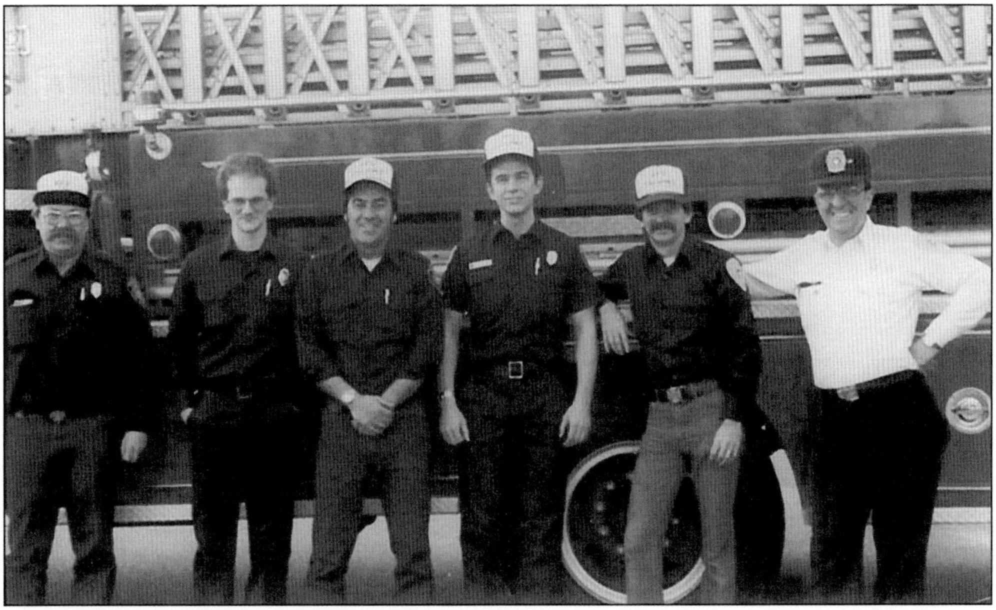

This early photograph of firefighters has them lined up at headquarters. They are, from left to right, William Callahan, Scott Carpenter, Raymond Perry, Steven Yetman, George Creamer, and Kenneth Mills. (Courtesy of Albert Fay.)

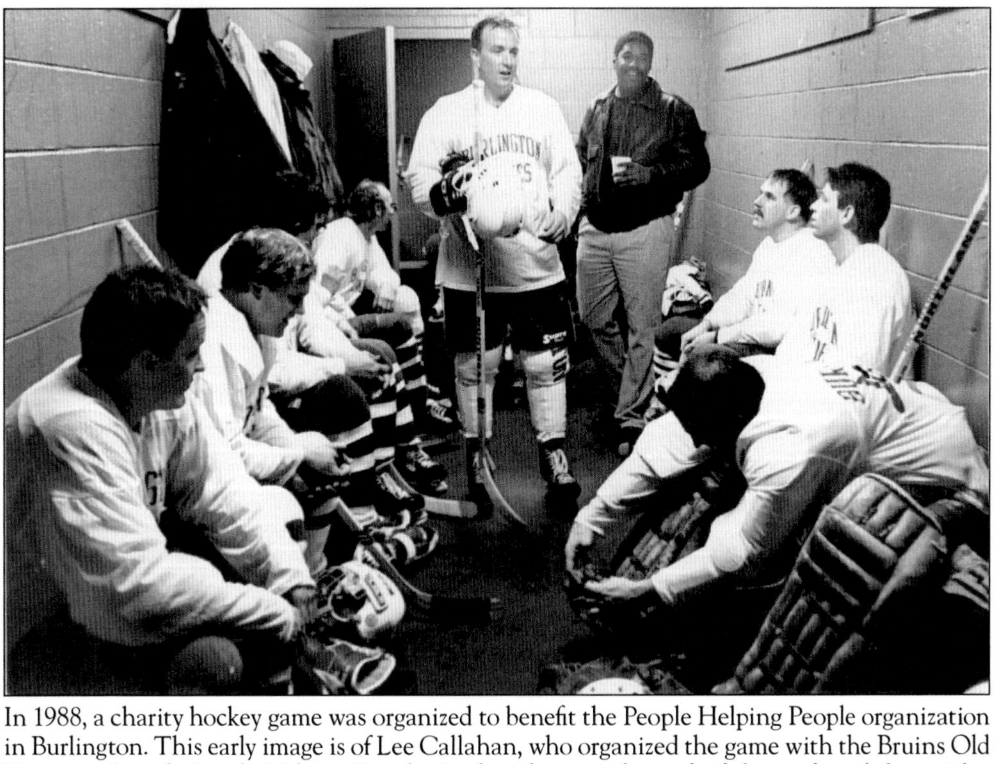

In 1988, a charity hockey game was organized to benefit the People Helping People organization in Burlington. This early image is of Lee Callahan, who organized the game with the Bruins Old Timers to benefit People Helping People. In this photograph are firefighters, from left to right, unidentified, Timothy Browne, unidentified, Gary Cauldwell, Lee Callahan, coach Augusto Grace, Gary Arbing, Jim Sherman, and John Corbett. (Courtesy of Albert Fay.)

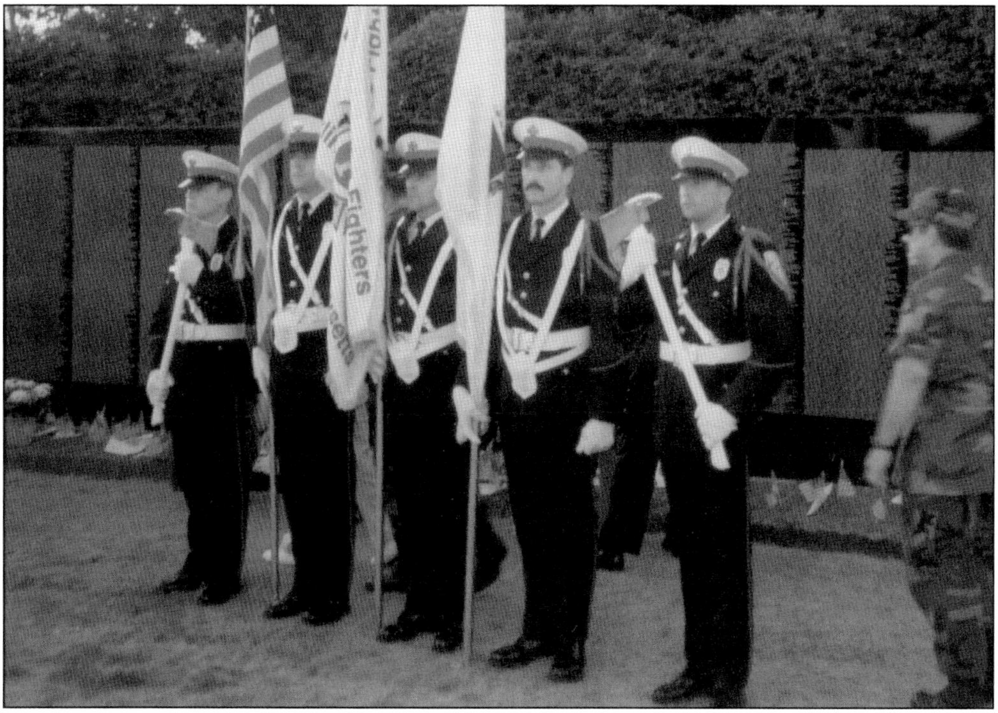

In 1999, the Burlington Fire Department color guard was present at the Vietnam Veterans Moving Wall Memorial visit to Burlington. The memorial was in honor of the men and women of the armed forces who gave their lives or who remain missing in the Vietnam War. (Courtesy of Paul Thibault.)

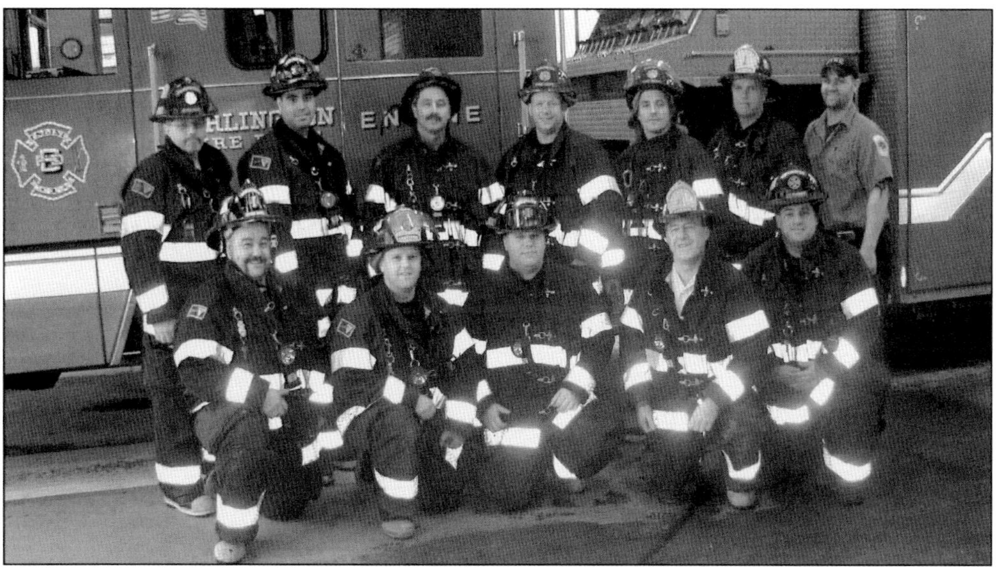

This is group four at headquarters. They are, from left to right, (first row) Clifford Comeau, Timothy Browne, Jeffrey Boucher, Capt. Mike Nolan, and David Angelo; (second row) Edward McLean, Gus Marino, David Sullivan, Sean Connors, Mark Cedrone, Lt. James Browne, and dispatcher Stephen Baia. Missing are Leonard Sawyer and Timothy Hovasse. (Courtesy of Burlington Fire Department.)

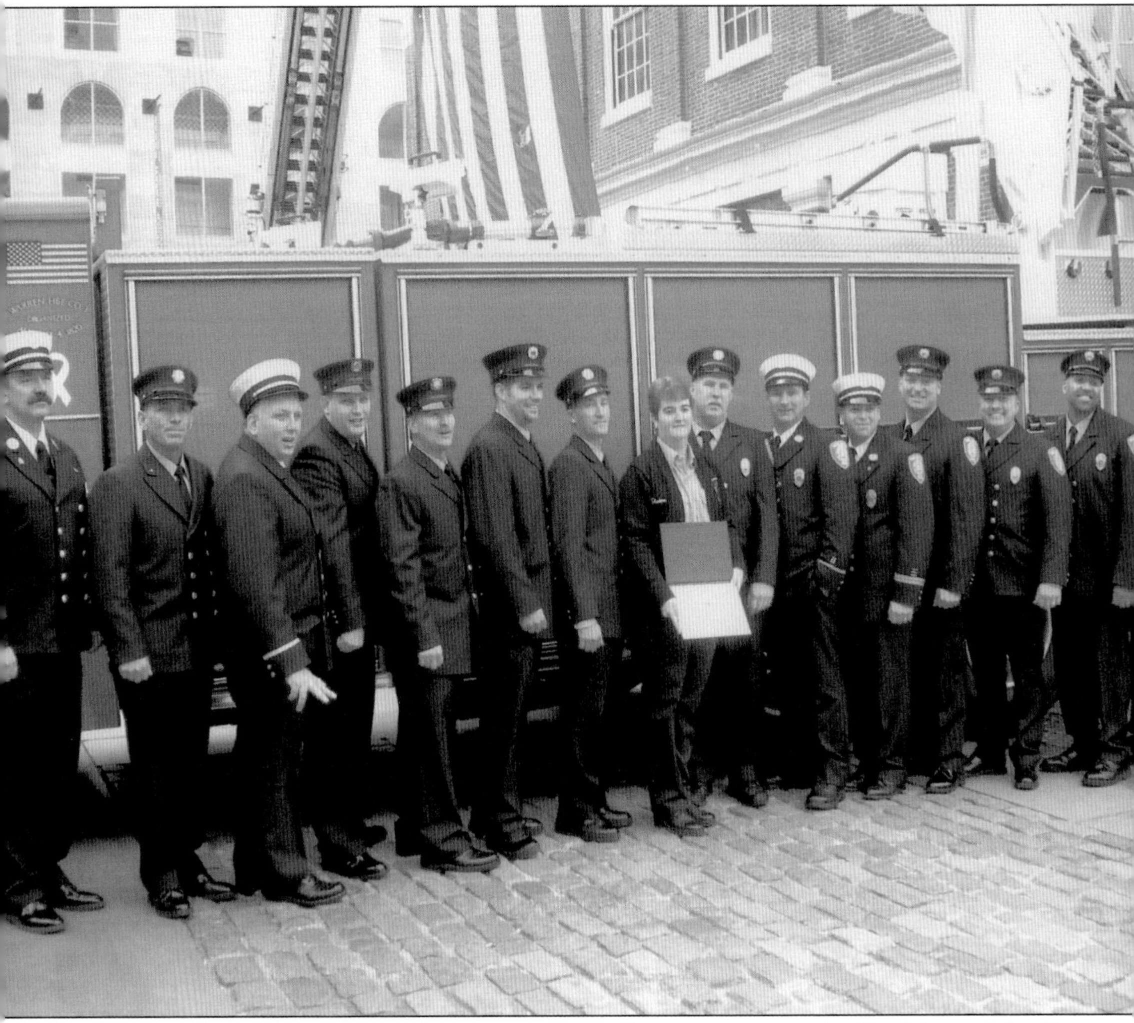

In 2005, group two was invited to a ceremony in Boston to receive the Firefighter of the Year Award. The firefighters of group two were called to a house fire and rescued former call firefighter Stanley Cooper from his Stoney Brook Road home. He later succumbed to the injuries he received in the fire. From left to right are Assistant Chief Steven Yetman, firefighter James Sherman, Lt. Peter McAnsepie, firefighters Ernest Covino, John Price, Brian Petrone, and Michael Fontannay, dispatcher Elaine Carpenter, firefighter Gary Arbing, Lt. Stephen McLean, Capt. Kevin Browne, firefighters Michael Kerrigan, Donald McDonald, and Frederick Williams, and Chief Lee Callahan. (Courtesy of Burlington Fire Department.)

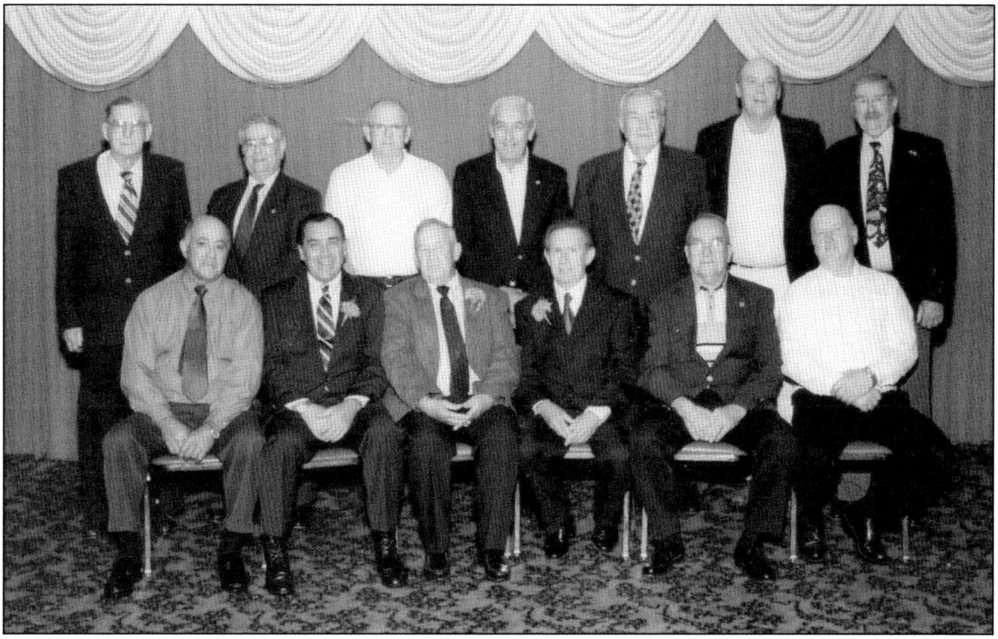

A retirement party in 2000 was hosted by the following firefighters, from left to right: (first row) Robert Guidoboni, Robert Perry, James Hapenney, James Fay, former chief Herbert Crawford, and Frederick Keene; (second row) Louis Skelton, Albert Fay, George Clark, Robert Sullivan, Walter Mountain, Douglas Hyde, and Walter DeCost. (Courtesy of Albert Fay.)

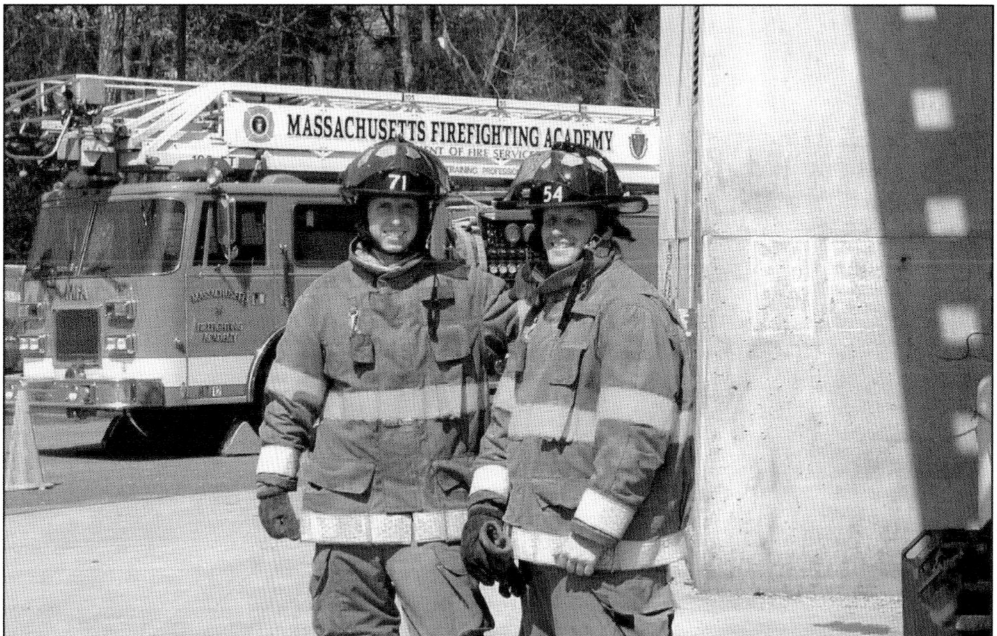

Firefighters Shaun Kenney (left) and Ernest Covino are shown at the Massachusetts Fire Academy. Covino created the fire department Web site. The Web site is an asset to the department, keeping the public aware of the department's role in the community. (Courtesy of Burlington Fire Department.)

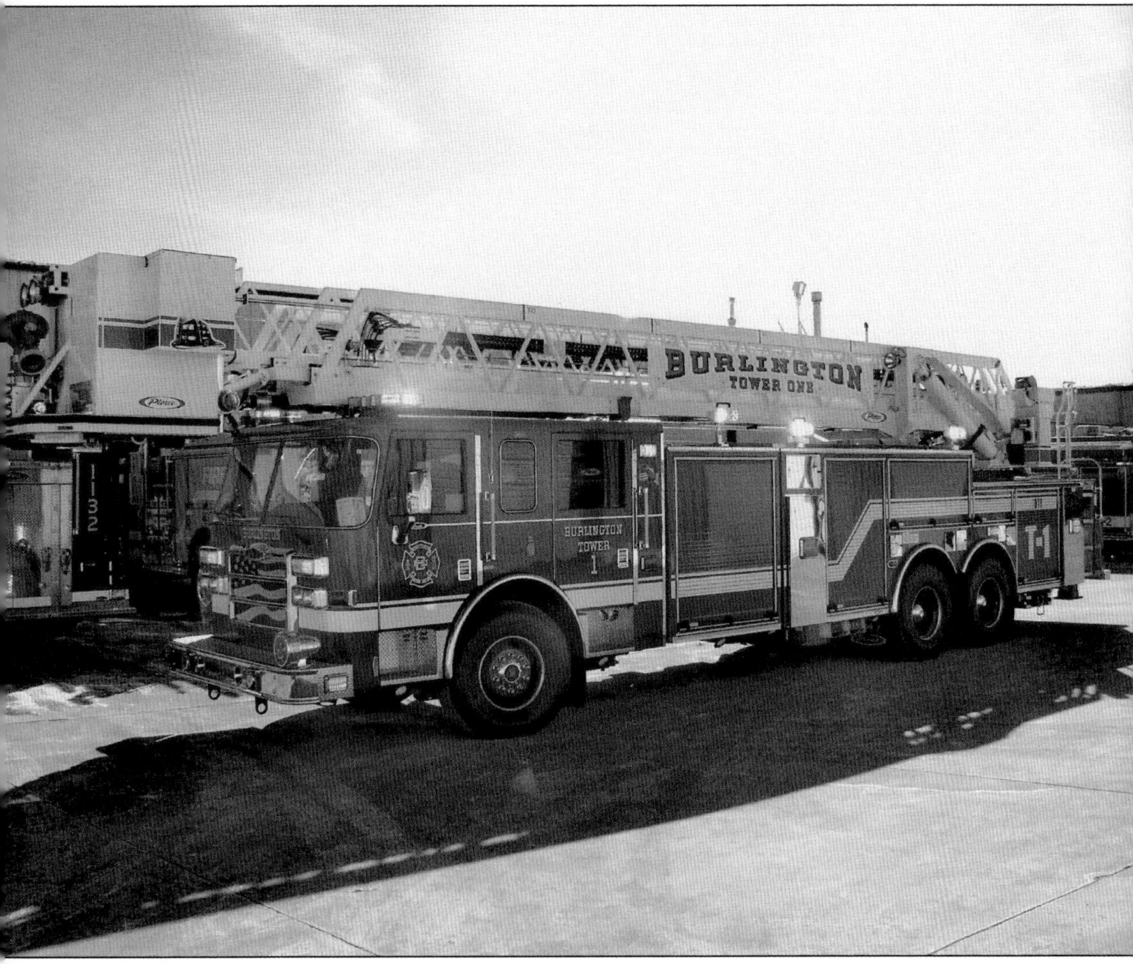

The Burlington Fire Department welcomed with pride a new 2007 Pierce 100-foot tower in March 2008. The cab is the Pierce Dash. Included is a Detroit Diesel Series 60 engine, independent front suspension, rollover protection, antilock brakes, 10,000-watt Harrison generator, pullout tool boards, two electrical reels with junction box, steel-framed ladder and basket, and master stream capable of flowing 2,000 gallons per minute. It replaced the 1990 tower, which was in service for 17 years. Firefighting equipment has come a long way from 1909, when the town bought a wagon and put 25 milk cans filled with water on it to fight fires. (Courtesy of Burlington Fire Department.)

BIBLIOGRAPHY

Corbett, Thomas. *The Burlington Firefighters*. Burlington, MA: Burlington Relief Association, Town of Burlington, and Local 2313, 1980.

Costa, Robert. *Burlington*. Charleston, SC: Arcadia Publishing, 2001.

Dunham, Lotta Cavanaugh Rice. *The History of Burlington 1640–1950*. Edited by Robert Zahora. 1998.

Faria, Lawrence. *Protective Breathing Apparatus*. Bowie, MD: Robert J. Brady Company, Prentice-Hall Company, 1975.

Fogelberg, John E. *Burlington, Part of a Greater Chronicle*. Burlington, MA: Burlington Historical Commission, 1976.

———. *Burlington Past and Present*. Compiled by James Ashworth. CD-ROM. 2005.

Goff, John. *1998–1999 Historic Preservation Survey of Burlington*. Burlington, MA: Town of Burlington, 1999.

Patterson, Capt. Michael. *Burlington Fire Department Training*. CD-ROM. 2007.

Town of Burlington, Massachusetts, Annual Report. Burlington, MA: 1902–2007.

Washburn, Robert. *History of the Burlington Fire Department*. Burlington, MA: Self-published, 1997.

ACROSS AMERICA, PEOPLE ARE DISCOVERING SOMETHING WONDERFUL. *THEIR HERITAGE.*

Arcadia Publishing is the leading local history publisher in the United States. With more than 3,000 titles in print and hundreds of new titles released every year, Arcadia has extensive specialized experience chronicling the history of communities and celebrating America's hidden stories, bringing to life the people, places, and events from the past. To discover the history of other communities across the nation, please visit:

www.arcadiapublishing.com

Customized search tools allow you to find regional history books about the town where you grew up, the cities where your friends and family live, the town where your parents met, or even that retirement spot you've been dreaming about.

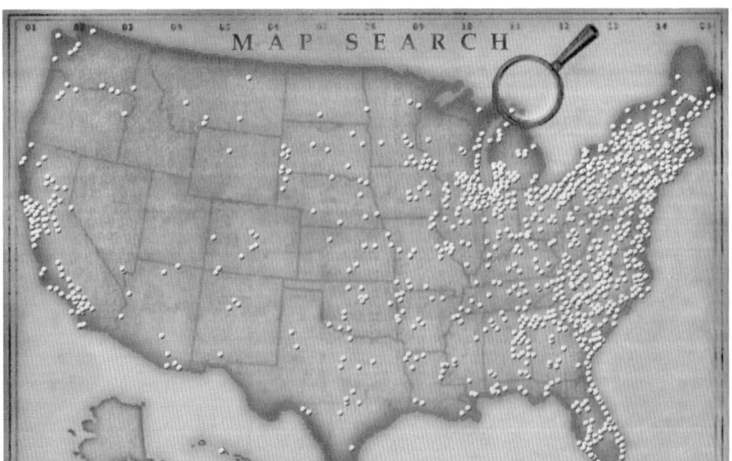